868TUTORS

Preparation for High School

Mathematics

Topic by Topic Solutions

Anim Adrian Amarsingh

Dedication

Dedicated to all ambitious

Students

and

Teachers

Introduction

In order to attain competence in Mathematics, students need to solve several Mathematical problems. Also, students need to obtain appropriate feedback on the questions that they have solved. This feedback is vital for progress. Moreover, the logical process is important when solving Mathematical problems. "Preparation for High School Mathematics: Topic by Topic Solutions" provides students and teachers with a wide variety of model answers for topics encountered in High School Mathematics. To obtain the maximum benefit of these workbook solutions, attempt the questions prior to reviewing the solutions. Additional worksheets and solutions are free for download on the companion website: www.868tutors.com. Please direct any questions and concerns to questions@868tutors.com.

TABLE OF CONTENTS

ALGEBRA	1
ALGEBRA II	16
BEARINGS	24
CIRCLE THEOREMS	33
COMPUTATION	50
COMSUMER ARITHMETIC	68
CONSUMER ARITHMETIC II	82
FUNCTIONS	92
INVESTIGATION	104
MATRICES	115
MEASUREMENT	136
MEASUREMENT II	143
MEASUREMENT III	152
QUADRATIC EQUATIONS	168
QUADRATIC GRAPHS	177
SETS	196
SIMULTANEOUS EQUATIONS	211
STATISTICS	221
STATISTICS II	233
STRAIGHT LINE GRAPHS	250
SUBJECT OF THE FORMULA	267
TRIGONOMETRY RIGHT-ANGLED	277
TRIGONOMETRY NON-RIGHT ANGLED	291
TRIGONOMETRY COMBINED	305
VECTORS	314

868 TUTORS

Preparation for
High School Mathematics
Algebra
Solutions

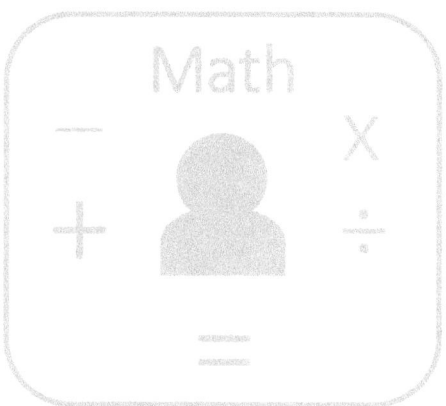

Instructions and Tips:

- ✓ You have 90 minutes to complete this worksheet
- ✓ This worksheet consists of 13 questions
- ✓ Write answers in the spaces provided
- ✓ All working must be clearly shown

Student Name: _____

Student ID: _____

Date: __ / __ / ____

Total Score:

Highest Score:

Tutor's Comments:

Access more free worksheets at www.868tutors.com

Question 1

For each question below, perform the appropriate substitution and determine the final result.

m = 10   n = 2   p = 3

(a)   **5n + 2p** = 5(2) + 2(3) = 10 + 6 = $\boxed{16}$

(b)   **3m - 6p** = 3(10) − 6(3) = 30 − 18 = $\boxed{12}$

(c)   **2n + 3n + 2p** = 2(2) + 3(2) + 2(3) = 4 + 6 + 6 = $\boxed{16}$

(d)   **3m + 3p** = 3(10) + 3(3) = 30 + 9 = $\boxed{39}$

(e)   **6mnp** = 6(10) (2) (3) = 120(3) = $\boxed{360}$

(5 marks)

Question 2

For each question below, perform the appropriate substitution and determine the final result.

$x = 1$ $y = 2$ $z = 3$

(a) $2x + 2y + 2z$ = 2(1) + 2(2) + 2(3) = 2 + 4 + 6 = $\boxed{12}$

(b) $3x - 3y + z^3$ = 3(1) - 3(2) + (3)^3 = 3 − 6 + 27 = $\boxed{24}$

(c) $3z + 2x + 2y$ = 3(3) + 2(1) + 2(2) = 9 + 2 + 4 = 11 + 4 = $\boxed{15}$

(d) $4xzy^2$ = 4(1)(3)(2)^2 = 4 × 1 × 3 × 4 = $\boxed{48}$

(e) $6xyz$ = 6(1)(2)(3) = 6 × 1 × 2 × 3 = $\boxed{36}$

(5 marks)

Question 3

For each question below, perform the appropriate substitution and determine the final result.

q = 0   r = 1   s = 2

(a)   **2q + 2r +2s^q** = $2(0) + 2(1) + 2(2)^0$ = 0 + 2 + 2(1) = 2 + 2 = $\boxed{4}$

(b)   **3q + s³** = $3(0) + (2)^3$ = 0 + 8 = $\boxed{8}$

(c)   **3r + 2r³** = $3(1) + 2(1)^3$ = 3 + 2 = $\boxed{5}$

(d)   **4rqs²** = $4(1)(0)(2)^2$ = 4 × 1 × 0 × 4 = $\boxed{0}$

(e)   **4rs** = $4(1)(2)$ = 4 × 1 × 2 = $\boxed{8}$

(5 marks)

Question 4

For each question below, perform the appropriate substitution and determine the final result:

$t = 0 \quad u = 1 \quad v = 2$

(a) $\dfrac{2v}{u} = \dfrac{2(2)}{1} = \dfrac{4}{1} = \boxed{4}$

(b) $\dfrac{2t}{u} = \dfrac{2(0)}{1} = \boxed{0}$

(c) $\dfrac{(2v)^2}{u} = \dfrac{(2\times 2)^2}{1} = \dfrac{(4)^2}{1} = \dfrac{16}{1} = \boxed{16}$

(d) $\dfrac{2v^2}{u} = \dfrac{2\times (2)^2}{1} = \dfrac{8}{1} = \boxed{8}$

(4 marks)

Question 5

Simplify the following algebraic expressions:

(a) 3a + 6a + 2a = $\boxed{11a}$

(b) 2y + 11y + 3y = $\boxed{16y}$

(c) 3z + 2z - 1z = $\boxed{4z}$

(d) 5y + 10y - 2y = $\boxed{13y}$

(e) 6c + 2c - 1c = $\boxed{7c}$

(5 marks)

Question 6

Simplify the following algebraic expressions:

(a) $\dfrac{3c^2}{c} = \dfrac{3 \times c \times c}{c} = \dfrac{3 \times c}{1} = \dfrac{3c}{1} = \boxed{3c}$

(b) $\dfrac{9a^3}{9a} = \dfrac{9 \times a \times a \times a}{9a} = \boxed{a^2}$

(c) $\dfrac{10b^4}{2b} = \dfrac{10 \times b \times b \times b \times b}{2 \times b} = \dfrac{5 \times b \times b \times b}{1} = \boxed{5b^3}$

(d) $\dfrac{3c^2}{c^5} = \dfrac{3 \times c \times c}{c \times c \times c \times c \times c} = \boxed{\dfrac{3}{c^3}}$

(4 marks)

Question 7

(a)

One Snow Cone costs s dollars

One Pack of Kurma costs p dollars

Write an expression for the total cost of 7 snow cones and 5 packs of Kurma

$\boxed{7s + 5p}$

(b)

One Tamarind Ball costs x cents

One Salted Prune costs y cents

Write an expression for the total cost of 10 Tamarind Balls and 20 Salted Prunes

$\boxed{10x + 20y}$

(c)

One Soursop costs m cents

One Guava costs p cents

Write an expression for the total cost of 15 soursops and 12 guavas

$\boxed{15m + 12p}$

(3 marks)

Question 8

Simplify the following algebraic expressions:

(a) 3pq + 6mn + 2pq - 2mn

   3pq + 2pq + 6mn - 2mn

   $\boxed{5pq + 4mn}$

(b) 2bc + 1gr + 3bc - 6gr

   2bc + 3bc + 1gr - 6gr

   5bc - 5gr

   $\boxed{5(bc - gr)}$

(c) p × p × p × q × q × q = $\boxed{p^3 q^3}$

(d) 6b × 3q = $\boxed{18\ bq}$

(4 marks)

Question 9

Make the letter indicated in brackets the subject of the formula of the following formulae:

(a) **F = ma**   [a]
$$ma = F$$
$$am = F$$
$$\boxed{a = \frac{F}{m}}$$

(b) **v = u + at**   [u]
$$u + at = v$$
$$\boxed{u = v - at}$$

(c) **M = frs**   [r]
$$frs = M$$
$$r \times f \times s = M$$
$$\boxed{r = \frac{M}{fs}}$$

(d) **g = ft**   [t]
$$t \times f = g$$
$$\boxed{t = \frac{g}{f}}$$

(4 marks)

Question 10

Make the letter indicated in brackets the subject of the formula of the following formulae:

(a) $Y = \dfrac{10b^2}{zh}$ [b]

$\dfrac{Y}{1} = \dfrac{10b^2}{zh}$ (Cross-multiplying)

$Yzh = 10b^2 \quad 10b^2 = Yzh \quad b^2 = \dfrac{Yzh}{10}$

$\boxed{b = \sqrt{\dfrac{Yzh}{10}}}$

(b) $S = \dfrac{10c^2}{d^2}$ [d]

$\dfrac{S}{1} = \dfrac{10c^2}{d^2}$ (Cross-multiplying)

$10c^2 = Sd^2 \quad Sd^2 = 10c^2$

$d^2 = \dfrac{10c^2}{S} \quad \boxed{d = \sqrt{\dfrac{10c^2}{S}}}$

(c) $S = \dfrac{10c^3}{d^2}$ [c]

$\dfrac{S}{1} = \dfrac{10c^3}{d^2}$ (Cross-multiplying)

$10c^3 = Sd^2$

$c^3 = \left(\dfrac{Sd^2}{10}\right)$

$\boxed{c = \sqrt[3]{\dfrac{Sd^2}{10}}}$

(6 marks)

Question 11

Make the letter indicated in brackets the subject of the formula of the following formulae:

(a) $x = qr^4$ [q]

$qr^4 = x$

$\boxed{q = \dfrac{x}{r^4}}$

(b) $g = u^2 + vl$ [u]

$u^2 + vl = g$

$u^2 = g - vl$

$\boxed{u = \sqrt{g - vl}}$

(c) $M^2 = frs$ [M]

$\boxed{M = \sqrt{frs}}$

(d) $k = ft^2$ [t]

$t^2 f = k$

$t^2 = \dfrac{k}{f}$

$\boxed{t = \sqrt{\dfrac{k}{f}}}$

(8 marks)

Question 12

Factorize the following:

(a) $3y^2 + 6y$

$\boxed{3y(y+2)}$

(b) $10c^3 + 100c$

$\boxed{10c(c^2 + 10)}$

(c) $5d^2 + 10d^3 + 15d$

$\boxed{5d(d + 2d^2 + 3)}$

(3 marks)

Question 13

Expand the following:

(a) $3(a + b) = \boxed{3a + 3b}$

(b) $5(y + z) = \boxed{5y + 5z}$

(c) $4(t + u + v) = \boxed{4t + 4u + 4v}$

(d) $3(d + a + c) = \boxed{3d + 3a + 3c}$

(e) $4(g + h) = \boxed{4g + 4h}$

(5 marks)

END OF WORKSHEET

868 TUTORS

Preparation for

High School Mathematics

Algebra II

Solutions

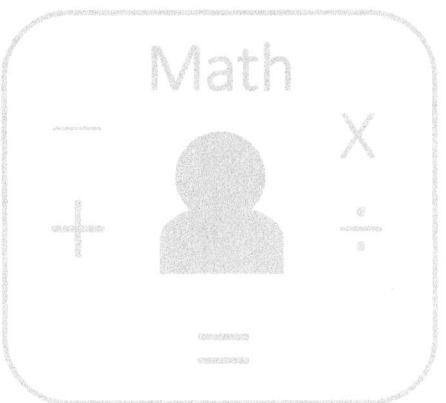

Instructions and Tips:

- ✓ You have 40 minutes to complete this worksheet
- ✓ This worksheet consists of 6 questions
- ✓ Write answers in the spaces provided

Student Name: _____

Student ID: _____

Date: __/__/____

Total Score:

Highest Score:

Tutor's Comments:

Access more free worksheets at www.868tutors.com

Question 1

Find the product of the following binomial expressions:

(a) $(x+1)(x+2) = x \times x + x \times 2 + 1 \times x + 1 \times 2 =$
$x^2 + 2x + 1x + 2 =$
$\boxed{x^2 + 3x + 2}$

(b) $(x-1)(x-3) = x \times x + x \times -3 + -1 \times x + -1 \times -3$
$= x^2 - 3x - 1x + 3$
$= \boxed{x^2 - 4x + 3}$

(c) $(x+6)(x+5) = x \times x + x \times 5 + 6 \times x + 6 \times 5$
$= x^2 + 5x + 6x + 30$
$= \boxed{x^2 + 11x + 30}$

(d) $(x+2)(x+8) = x \times x + x \times 8 + 2 \times x + 2 \times 8$
$= x^2 + 8x + 2x + 16$
$= \boxed{x^2 + 10x + 16}$

(e) $(x-9)(x+4) = x \times x + x \times 4 + -9 \times x + -9 \times 4$
$= x^2 + 4x - 9x - 36$
$= \boxed{x^2 - 5x - 36}$

(f) $(-x+1)(x+2) = -x \times x + -x \times 2 + 1 \times x + 1 \times 2$
$= -x^2 + -2x + 1x + 2$
$= \boxed{-x^2 - x + 2}$

(g) $(x+10)(x-10) = x \times x + x \times -10 + 10 \times x + 10 \times (-10)$
$= x^2 + -10x + 10x - 100$
$\boxed{x^2 - 100}$

(7 marks)

Question 2

Find the product of the following binomial expressions:

(a) (2x+4) (x-3) = $2x \times x + 2x \times -3 + 4 \times x + 4 \times -3$

$\qquad = 2x^2 - 6x + 4x - 12$

$\qquad = \boxed{2x^2 - 2x - 12}$

(b) (2x+3) (2x-3) = $2x \times 2x + 2x \times -3 + 3 \times 2x + 3 \times -3$

$\qquad = 4x^2 - 6x + 6x - 9$

$\qquad = \boxed{4x^2 - 9}$

(c) (3x+1) (2x+4) = $3x \times 2x + 3x \times 4 + 1 \times 2x + 1 \times 4$

$\qquad = 6x^2 + 12x + 2x + 4$

$\qquad = \boxed{6x^2 + 14x + 4}$

(d) (2x+5) (x+3) = $2x \times x + 2x \times 3 + 5 \times x + 5 \times 3$

$\qquad = 2x^2 + 6x + 5x + 15$

$\qquad = \boxed{2x^2 + 11x + 15}$

(e) (4x+1) (x+6) = $4x \times x + 4x \times 6 + 1 \times x + 1 \times 6$

$\qquad = 4x^2 + 24x + 1x + 6$

$\qquad = \boxed{4x^2 + 25x + 6}$

(f) (3x+8) (x-4) = $3x \times x + 3x \times -4 + 8 \times x + 8 \times -4$

$\qquad = 3x^2 - 12x + 8x - 32$

$\qquad = \boxed{3x^2 - 4x - 32}$

(g) (x+9) (2x+1) = $x \times 2x + x \times 1 + 9 \times 2x + 9 \times 1$

$\qquad = 2x^2 + x + 18x + 9$

$\qquad = \boxed{2x^2 + 19x + 9}$ (14 marks)

Question 3

Factorise the following algebraic expressions:

(a) $3x^2 + 6xy^2$ = *3x (x + 2y²)*

(b) $2x^2y^3 + 4xy$ = *2xy (xy² + 2)*

(c) $3ab + 6a^2b^2 + 9a^3b^3$ = *3ab (1 + 2ab + 3a² b²)*

(d) $4ab + 8a^2b^2 + 16a^4b$ = *4ab (1 + 2ab + 4a³)*

(e) $3cd^3 + 9d^4$ = *3d³ (c + 3d)*

(f) $5m^2p + 10mnp + 20m^2$ = *5m (mp + 2np + 4m)*

(g) $a^2b^3 + a^4b^3 + ab$ = *ab (ab² + a³b² + 1)*

(7 marks)

Question 4

Factorise the following algebraic expressions:

(a) $a^3y + a^2y$ = $a^2y\,(a + 1)$

(b) $6pq + 12p^2q + 3p^4q^3$ = $3pq\,(2 + 4p + p^3q^2)$

(c) $a^2b^3 + ab + a^5b^5$ = $ab\,(ab^2 + 1 + a^4b^4)$

(d) $p^3q^3z + pqz^3$ = $pqz\,(p^2q^2 + z^2)$

(e) $12b^3 + 144b$ = $12b\,(b^2 + 12)$

(f) $3b^2 + 27b^4$ = $3b^2\,(1 + 9b^2)$

(g) $4km^2 + 8k^4m^3$ = $4km^2\,(1 + 2k^3m)$

(7 marks)

Question 5

Factorise the following:

(a) $x^2 + 3x + 2 = (x + 2)(x + 1)$

(b) $x^2 + 1x - 2 = (x + 2)(x - 1)$

(c) $x^2 + 7x + 10 = (x + 5)(x + 2)$

(d) $x^2 + 9x + 20 = (x + 5)(x + 4)$

(e) $x^2 - 2x - 24 = (x - 6)(x + 4)$

(f) $x^2 + 10x + 16 = (x + 8)(x + 2)$

(g) $x^2 + 7x - 8 = (x + 8)(x - 1)$

(7 marks)

Question 6

Factorise the following:

(a) $2x^2 + 7x + 3 =$ *(2x + 1) (x + 3)*

(b) $2x^2 + 8x + 8 =$ *(2x + 4) (x + 2)*

(c) $3x^2 - x - 2 =$ *(3x + 2) (x - 1)*

(d) $3x^2 + 5x + 2 =$ *(3x + 2) (x + 1)*

(8 marks)

END OF WORKSHEET

868 TUTORS

Preparation for

High School Mathematics

Bearings

Solutions

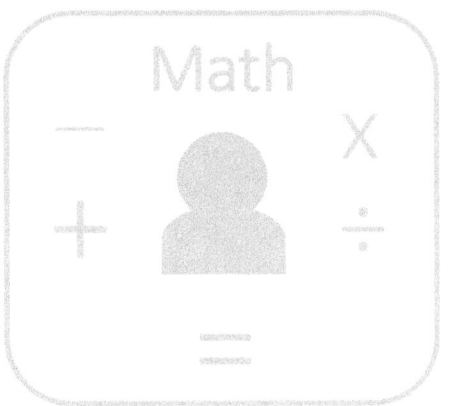

Instructions and Tips:

- ✓ You have 60 minutes to complete this worksheet
- ✓ This worksheet consists of 4 questions
- ✓ Write answers in the spaces provided
- ✓ Show all working

Student Name: _____

Student ID: _____

Date: __ / __ / ____

Total Score:

Highest Score:

Tutor's Comments:

Access more free worksheets at www.868tutors.com

Question 1

A ship leaves a port A and sails to an offshore oil platform 80 km away on a bearing of 070°. At port B, the ship changes course and sails to another port C, 50 km away on a bearing of 300°.

(a) Sketch the ship's journey and clearly indicate the following:
(i)   The direction of North
(ii) The points A, B and C
(iii) The bearings 070° and 300°
(iv) The distances 80 km and 50 km

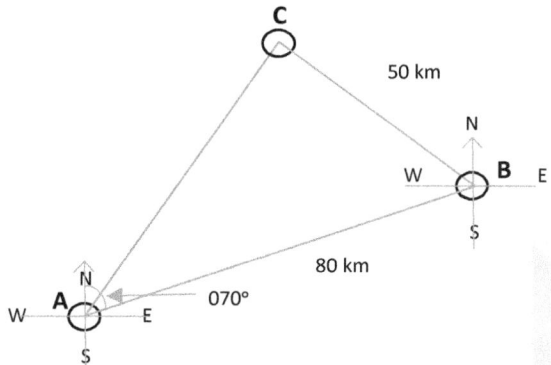

(4 marks)

(b) Calculate the straight line distance AC, in km, to 2 decimal places.

Using a simplified triangle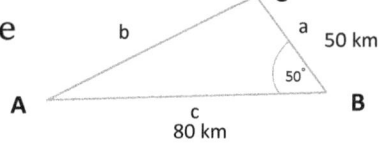

Applying cosine rule

$b^2 = a^2 + c^2 - 2ac \times \cos B$

$b^2 = (50)^2 + (80)^2 - 2(50)(80) \times \cos 50°$

$b^2 = 2500 + 6400 - 8000 \times \cos 50°$

$b^2 = 8900 - 5142.300877$

$b^2 = 3757.699123$ $b = 61.30$ km

$\boxed{AC = 61.30 \text{ km}}$

(3 marks)

(c) Calculate the bearing of C from A, to the nearest degree.

Using a simplified triangle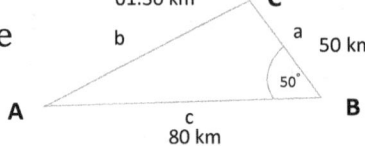

Applying Sine rule

$$\frac{a}{\sin A} = \frac{b}{\sin B} = \frac{c}{\sin C}$$

$$\frac{a}{\sin A} = \frac{b}{\sin B} \quad \frac{50 \text{ km}}{\sin A} \bowtie \frac{61.30 \text{ km}}{\sin 50°} \quad \text{(cross –multiplying)}$$

$50 \sin 50° = 61.30 \sin A$

$61.30 \sin A = 50 \sin 50°$

$\sin A = \frac{50 \sin 50°}{61.30}$

$\sin A = \frac{50 \sin 50°}{61.30}$

$\sin A = 0.624832335$

$A = \sin^{-1}(0.624832335)$

A = 38.67°

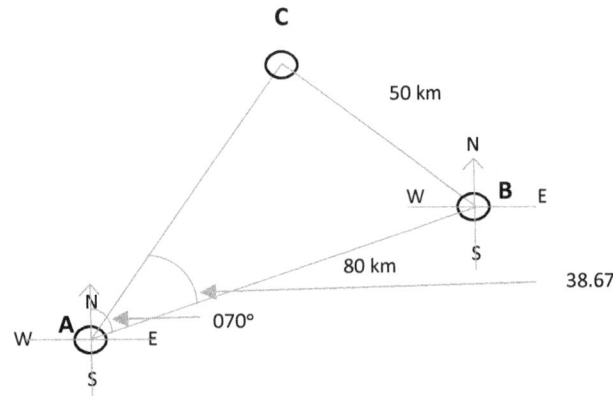

Bearing of C from A = 70° - 38.67°

Bearing of C from A = 31.33°

Bearing of C from A = 031°

(3 marks)

Question 2

A speedboat leaves a harbor J and heads to an islet, K, 30 km away on a bearing of 050°. At the islet K, the speedboat changes direction and heads to a port L, 40 km away on a bearing of 100°.

(a) Sketch the speedboat's journey and clearly indicate the following:
(i)   The direction of North
(ii)   The points J, K and L
(iii)   The bearings 050° and 100°
(iv)   The distances 30 km and 40 km

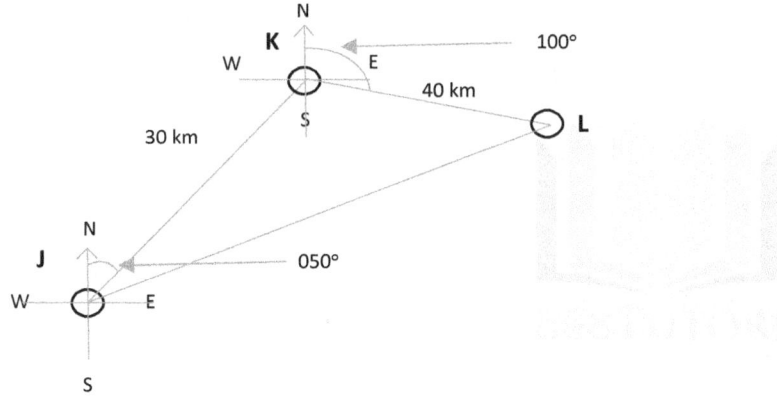

(4 marks)

(b) Calculate the straight line distance JL, in km, to 2 decimal places.

Using a simplified triangle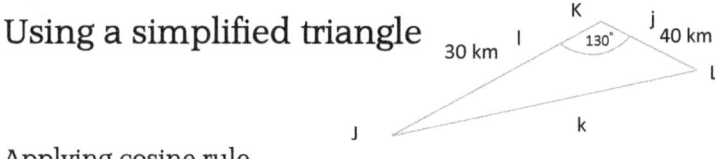

Applying cosine rule

$k^2 = j^2 + l^2 - 2jl \times \cos K$

$k^2 = (40)^2 + (30)^2 - 2(40)(30) \times \cos 130°$

$k^2 = 1600 + 900 - 2400 \times \cos 130°$

$k^2 = 2500 - -1542.690263$

$k^2 = 4042.690263$ $k = 63.58$ km   JL = 63.58 km

(3 marks)

(c) Calculate the bearing of L from J, to the nearest degree.

Using a simplified triangle

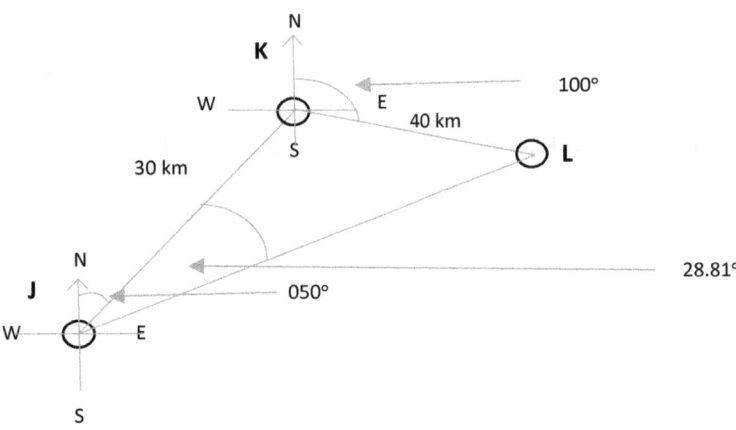

First we need to solve for J

Applying Sine rule $\quad \dfrac{j}{\sin J} = \dfrac{k}{\sin K} = \dfrac{l}{\sin L}$

$\dfrac{j}{\sin J} = \dfrac{k}{\sin K}$

$\dfrac{40}{\sin J} = \dfrac{63.58}{\sin 130°}$ (Cross-multiplying)

$40 \sin 130° = 63.58 \sin J$

$63.58 \sin J = 40 \sin 130°$

$\sin J = \dfrac{40 \sin 130°}{63.58}$

$\sin J = 0.481940511$

$J = \sin^{-1}(0.481940511) \quad J = 28.81°$

Bearing of L from J = $050° + 28.81°$

Bearing of L from J = 079° (to nearest degree)

(3 marks)

Question 3

Three hunters are positioned strategically in the Moruga forest. Hunter B is 50 meters north of Hunter A. Hunter C is on a bearing of 030° from Hunter A. The straight line distance between Hunter A and Hunter C is 38 m.

(a) Sketch the positions of the Hunters A, B and C
(i) The direction of North
(ii) The points A, B and C
(iii) The bearing 030°
(iv) The distances 50 m and 38 m

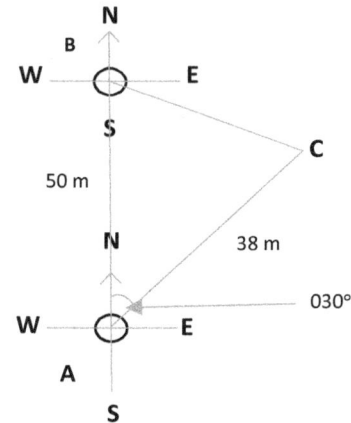

(4 marks)

(b) Calculate the straight line distance BC, in m, to 2 decimal places.

Using a simplified triangle

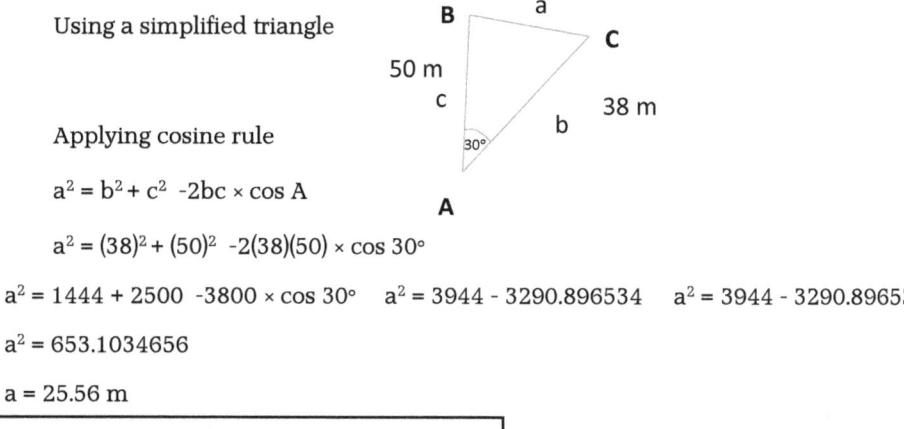

Applying cosine rule

$a^2 = b^2 + c^2 - 2bc \times \cos A$

$a^2 = (38)^2 + (50)^2 - 2(38)(50) \times \cos 30°$

$a^2 = 1444 + 2500 - 3800 \times \cos 30°$ $a^2 = 3944 - 3290.896534$ $a^2 = 3944 - 3290.896534$

$a^2 = 653.1034656$

$a = 25.56$ m

BC = 25.56 m (to 2 decimal places)

(3 marks)

(c) Calculate the bearing of C from B, to the nearest degree.

Using a simplified triangle

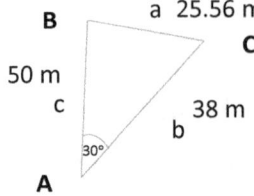

We need to solve for B

Applying sine rule

$$\frac{a}{\sin A} = \frac{b}{\sin B} = \frac{c}{\sin C}$$

$$\frac{a}{\sin A} = \frac{b}{\sin B}$$

$$\frac{25.56}{\sin 30°} = \frac{38}{\sin B}$$

38 sin 30° = 25.56 sin B

25.56 sin B = 38 sin 30°

$$\sin B = \frac{38 \sin 30°}{25.56}$$

$$\sin B = \frac{19}{25.56}$$

sin B = 0.743348982

B = sin^{-1} (0.743348982)

B = 48.02°

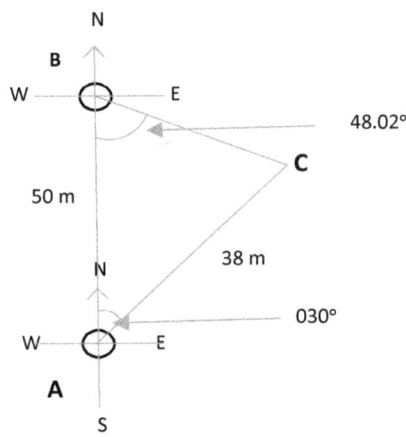

Bearing of C from B = 180° - 48.02°

Bearing of C from B = 132° (3 marks)

Question 4

The diagram below illustrates the position of three sea vessels in the Columbus Channel in the Southern Caribbean.

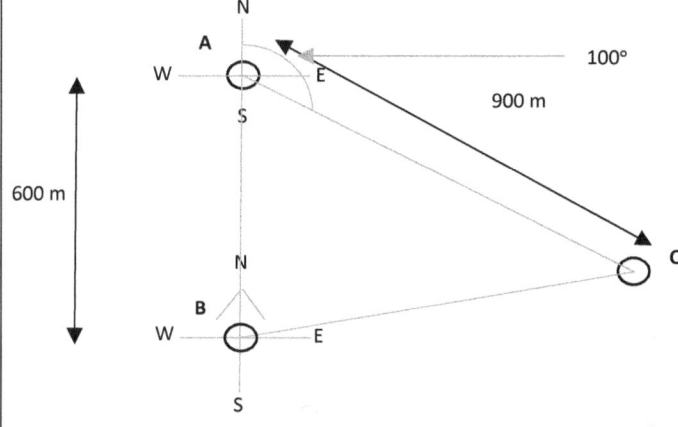

(Diagram not drawn to scale)

(a) Indicate the following on the diagram above:
(i) The bearing of vessel C from vessel A is 100°
(ii) The distance AC is 900 m
(iii) The distance BA is 600 m
(b) Calculate the straight line distance between B and C, to 2 decimal places.

Using a simplified triangle

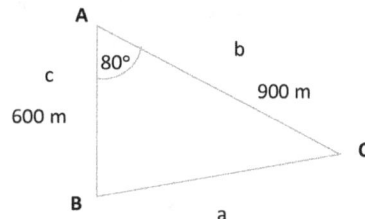

Applying cosine rule

$a^2 = b^2 + c^2 - 2bc \times \cos A$

$a^2 = (900)^2 + (600)^2 - 2(900)(600) \times \cos 80°$

$a^2 = 810{,}000 + 360{,}000 - 1{,}080{,}000 \times \cos 80°$

$a^2 = 1{,}170{,}000 - 187{,}540.0319$

$a^2 = 982{,}459.9681$

$a = 991.19$ m   **BC = 991.19 m (to 2 decimal places)**

(4 marks)

END OF WORKSHEET

868 TUTORS

Preparation for

High School Mathematics

Circle Theorems

Solutions

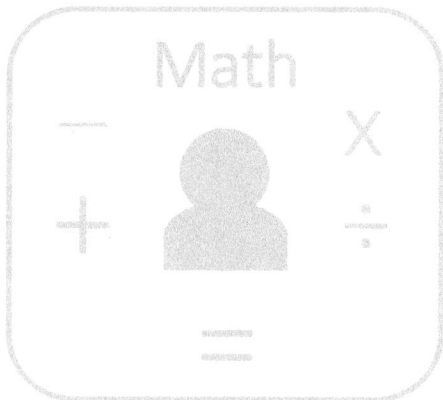

Instructions and Tips:

- You have 60 minutes to complete this worksheet
- This worksheet consists of 15 questions
- Write answers in the spaces provided
- All working must be clearly shown
- Diagrams are not drawn to scale

Student Name: _____

Student ID: _____

Date: __ /__ /____

Total Score:

Highest Score:

Tutor's Comments:

Access more free worksheets at www.868tutors.com

Question 1

Consider the circle below with centre Y:

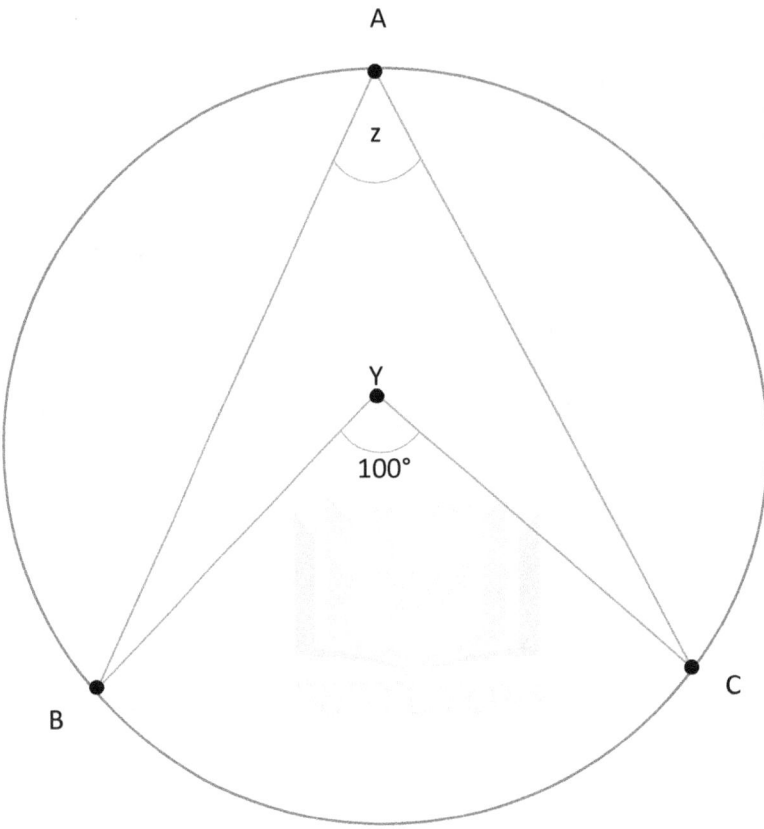

State the value of z and give a reason for your answer.

$Z = \frac{100°}{2} = 50°$ **Z = 50°**

Reason: The angle at the centre is twice the angle at the circumference.

(2 marks)

Question 2

Consider the circle below with centre D:

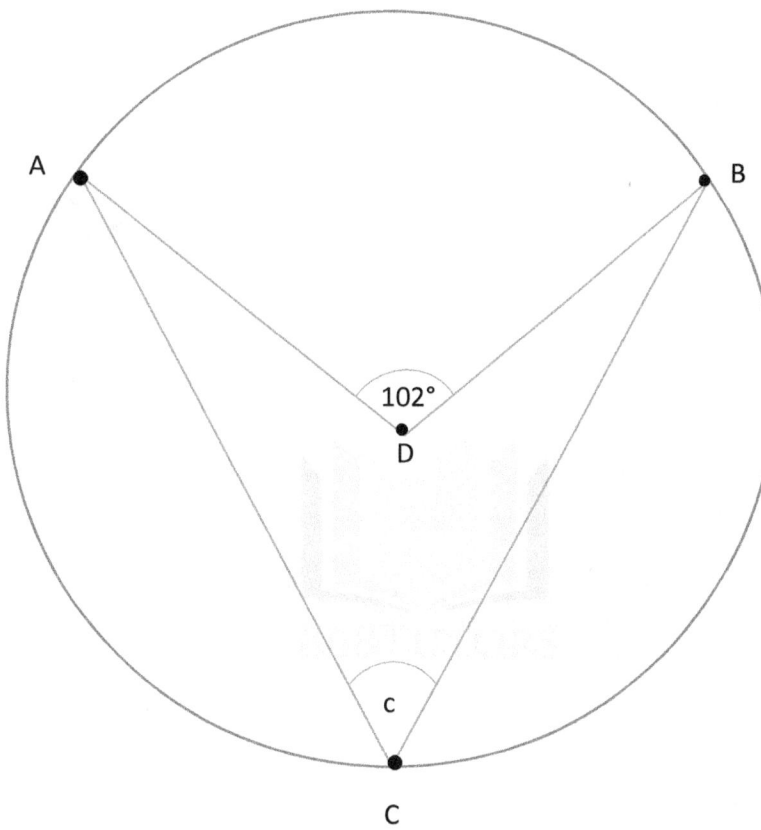

State the value of c and give a reason for your answer.

$c = \frac{102°}{2}$ **c = 51°**

Reason: The angle at the centre is twice the angle at the circumference.

(2 marks)

Question 3

Consider the circle below with centre P:

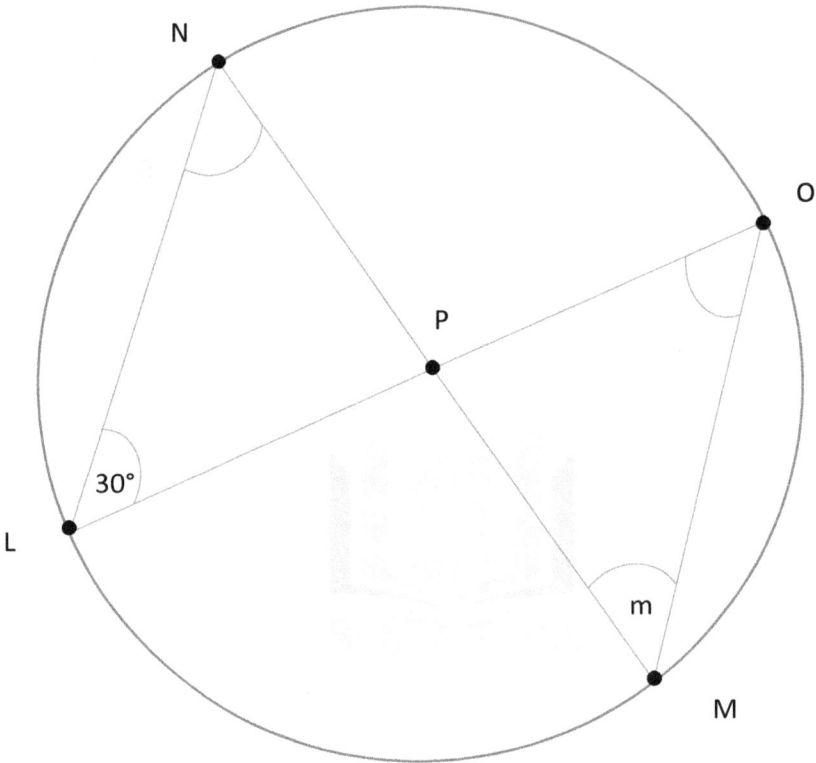

State the value of m and give a reason for your answer.

m = 30°

Reason: Angles in the same segment are equal.

(2 marks)

Question 4

Consider the circle below with centre V:

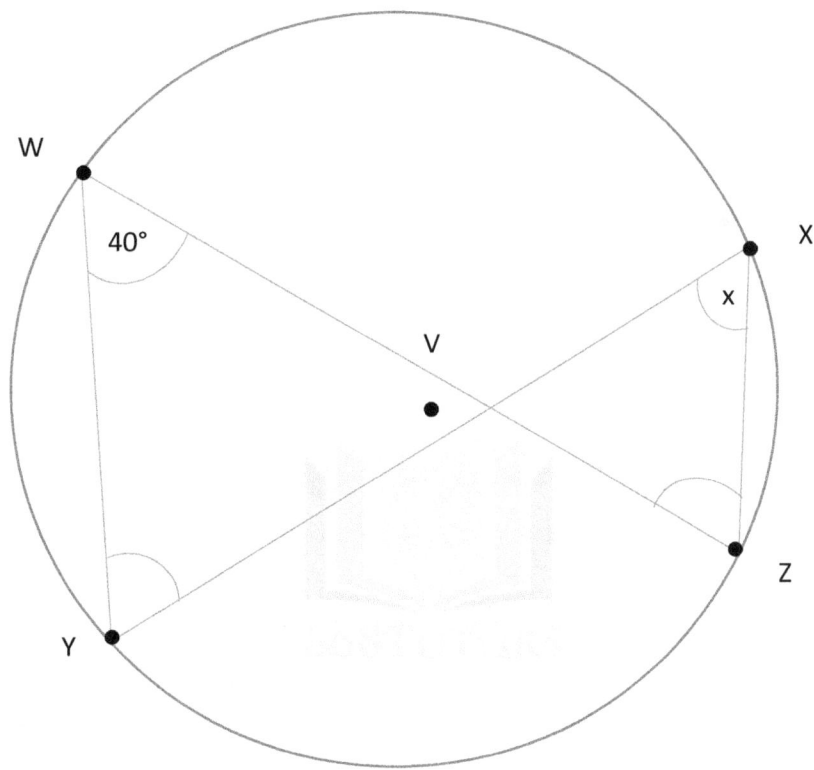

State the value of x and give a reason for your answer.

x = 40°

Reason: Angles in the same segment are equal.

(2 marks)

Question 5

Consider the circle below:

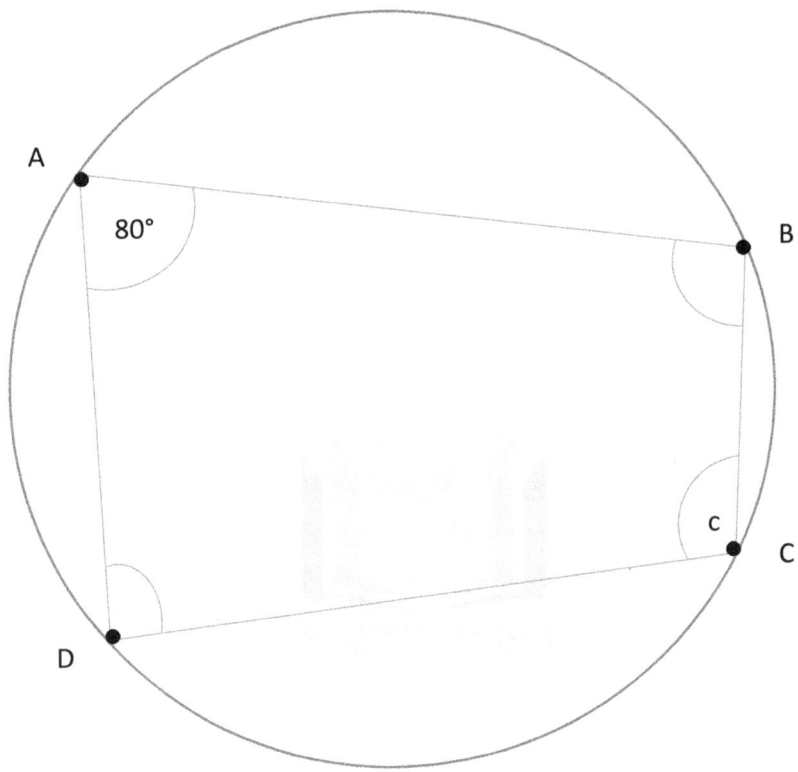

State the value of c and give a reason for your answer.

$c = 180° - 80°$ **c = 100°**

Reason: Opposite angles in a cyclic quadrilateral sum to 180°.

(2 marks)

Question 6

Consider the circle below:

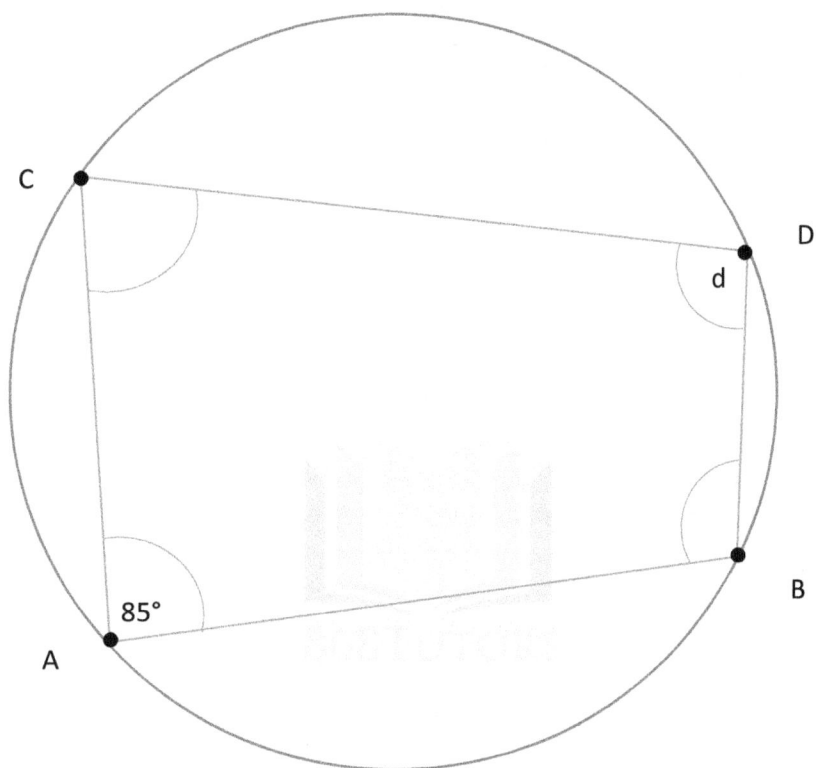

State the value of d and give a reason for your answer.

d = 180° - 85°

d = 95°

Reason: Opposite angles in a cyclic quadrilateral sum to 180°.

(2 marks)

Question 7

Consider the circle below:

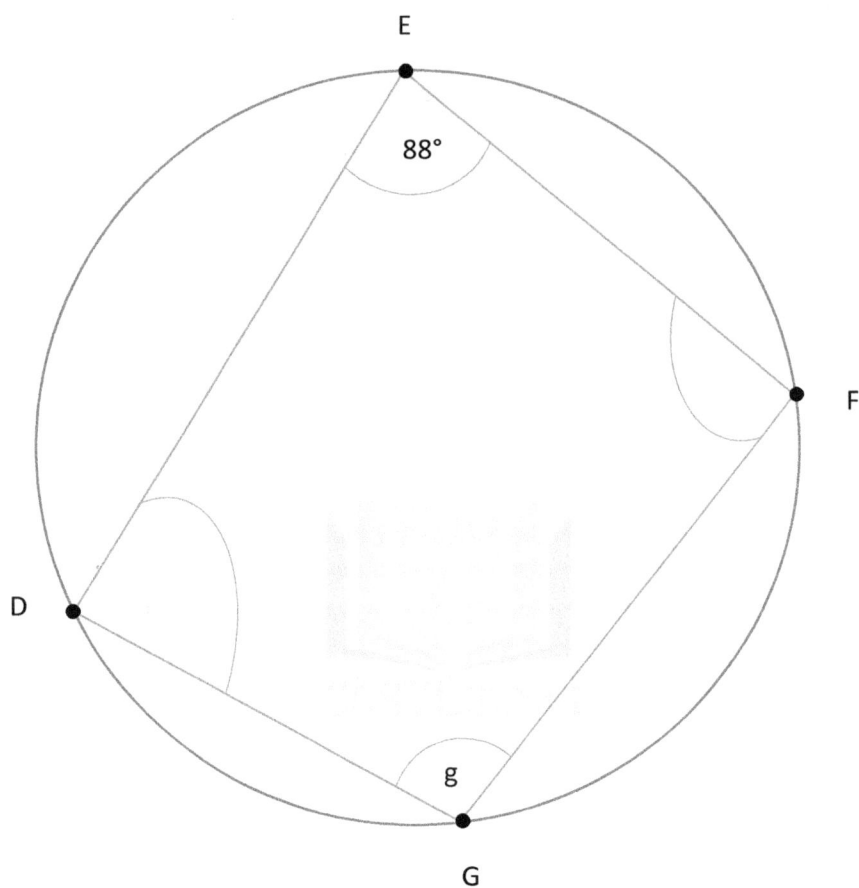

State the value of g and give a reason for your answer.

g = 180° - 88°

g = 92°

Reason: Opposite angles in a cyclic quadrilateral sum to 180°.

(2 marks)

Question 8

Consider the circle below with centre Z

Also, KL is a diameter of the circle below

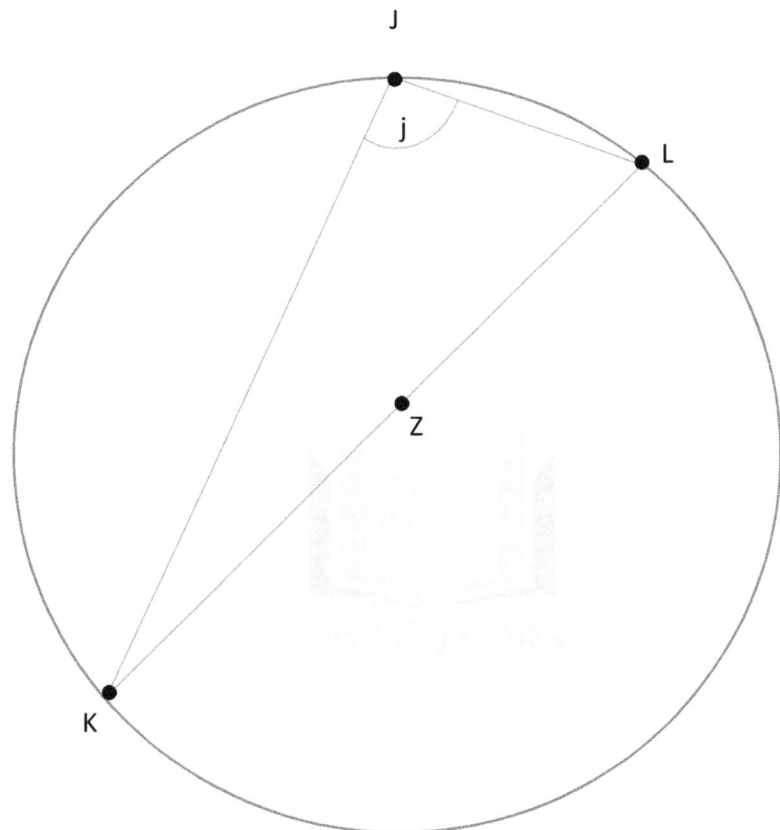

State the value of j and give a reason for your answer.

j = 90°

Reason: *The angle in a semi-circle is = 90°.*

(3 marks)

Question 9

Consider the circle below with centre D

Also, AB is a diameter of the circle below

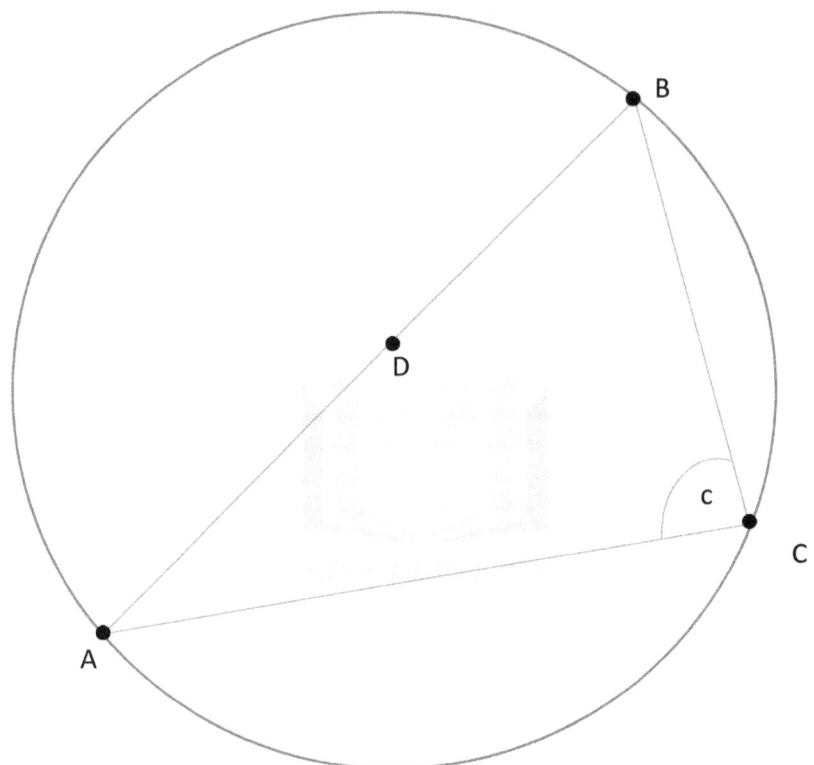

State the value of c and give a reason for your answer.

c = 90°

Reason: The angle in a semi-circle is = 90°.

(4 marks)

Question 10

Consider the circle below with centre Y:

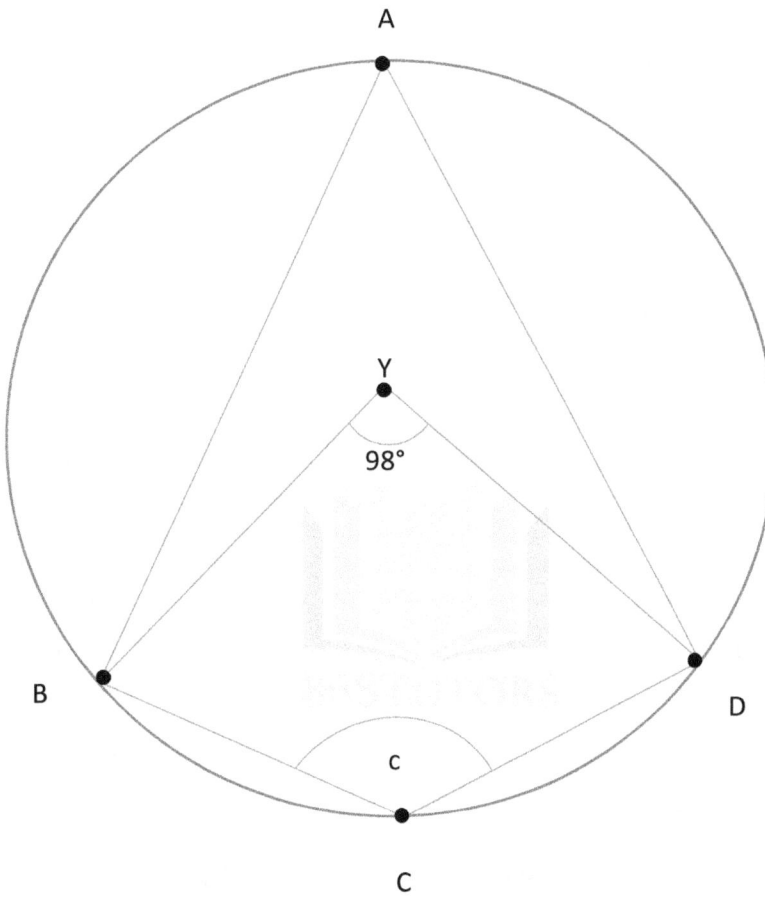

Determine the value of c and give reasons for your answer.

< DAB = $\frac{98°}{2}$ = 49°   Reason: The angle at the centre is twice the angle at the circumference

c = 180° - 49° (CBAD is a cyclic quadrilateral)

Reason: Opposite angles in a cyclic quadrilateral sum to 180°

c = 131°.

(5 marks)

Question 11

Consider the circle below with centre Y

XZ is a circle tangent and XY is a circle radius

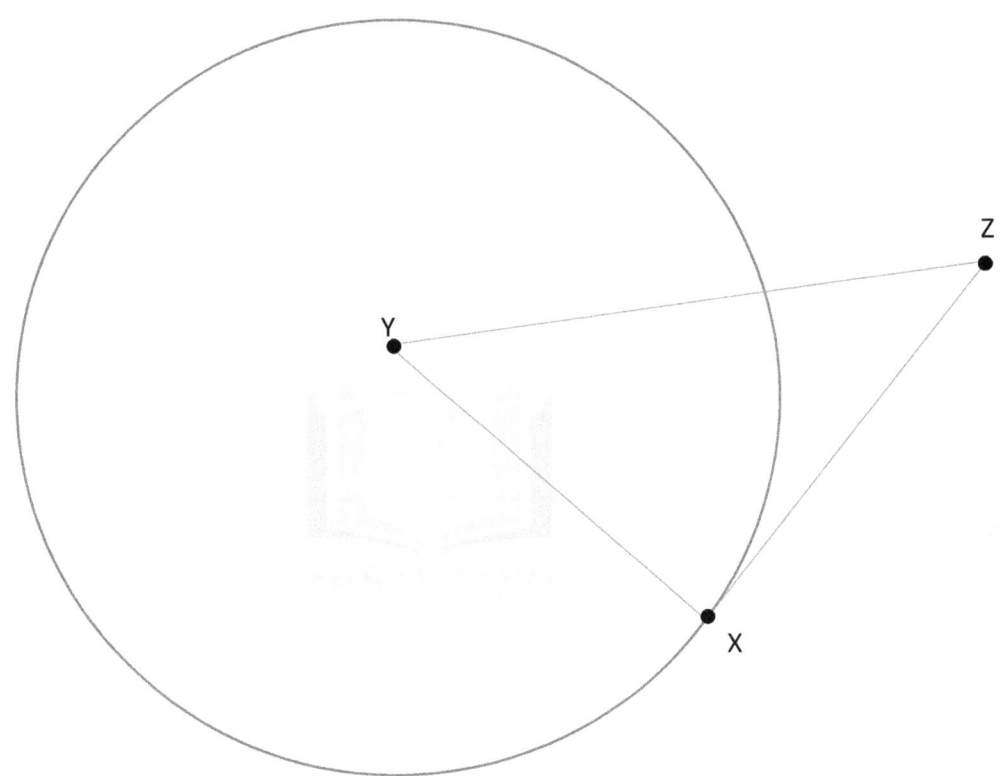

State the value of <YXZ and give a reason for your answer.

<YXZ = 90°

The angle between a circle tangent and a circle radius = 90°.

(3 marks)

Question 12

Consider the circle below with centre Y

XZ is a circle tangent and XY is a circle radius

<XZY = 30°

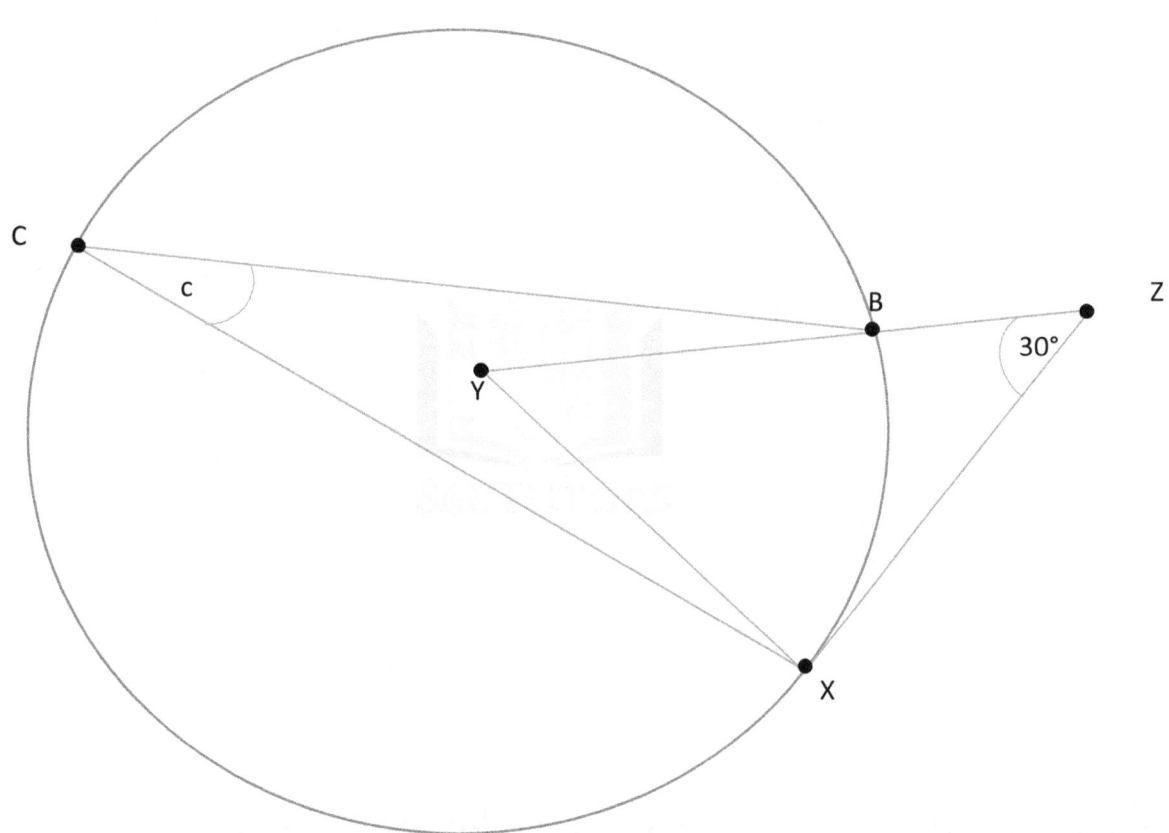

Determine the value of c and give reasons for your answer.

<YXZ = 90° (Reason: The angle between a circle tangent and circle radius = 90°.)

<XYZ = 180° - (90° + 30°)

<XYZ = 180° - 120° = 60° (Reason: Internal angles in a triangle sum to 180°.)

$c = \frac{60°}{2}$ **c = 30°** (Reason: The angle at the centre is twice the angle at the circumference.)

(3 marks)

Question 13

Consider the circle below with centre C

Also, AB is a diameter of the circle below

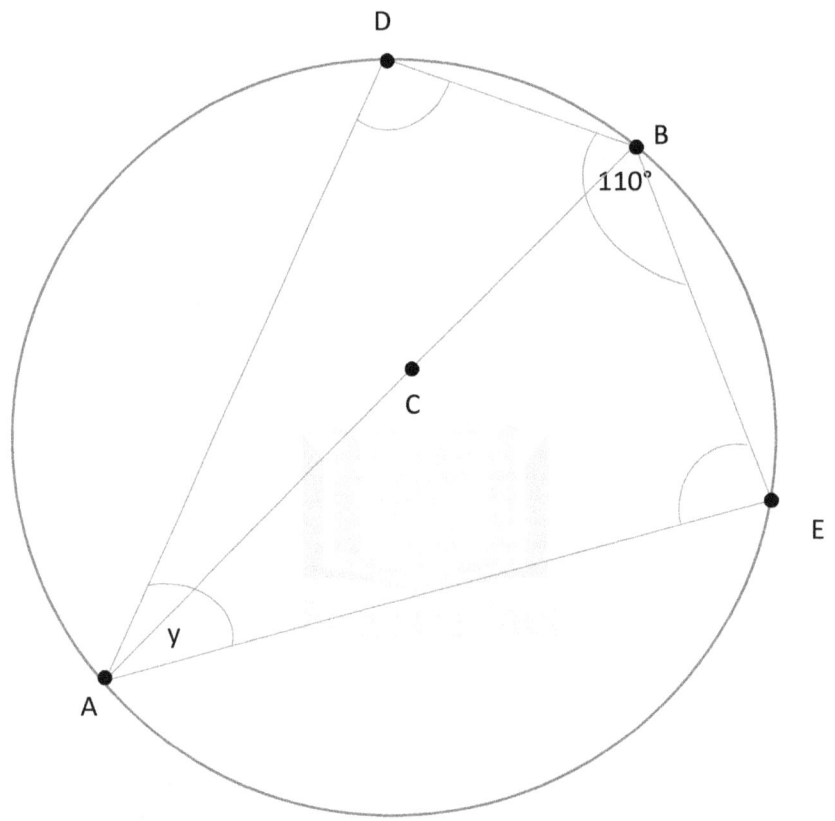

State the value of y and give a reason for your answer.

y = 180° - 110°

y = 70°

(Reason: Opposite angles in a cyclic quadrilateral sum to 180°.)

(6 marks)

Question 14

Consider the circle below with centre D

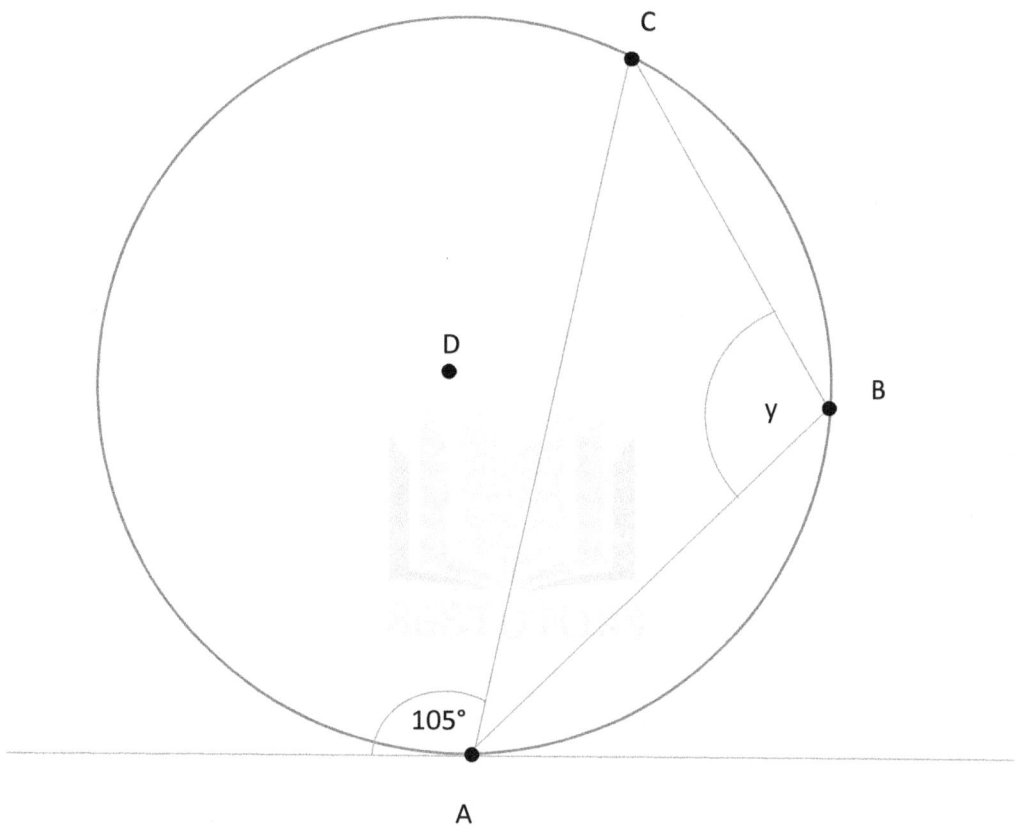

State the value of y and give a reason for your answer.

y = 105°

Reason: Alternate segment theorem

(3 marks)

Question 15

Consider the circle below with centre E

Line XY is a straight line that is tangent to the circle at point A

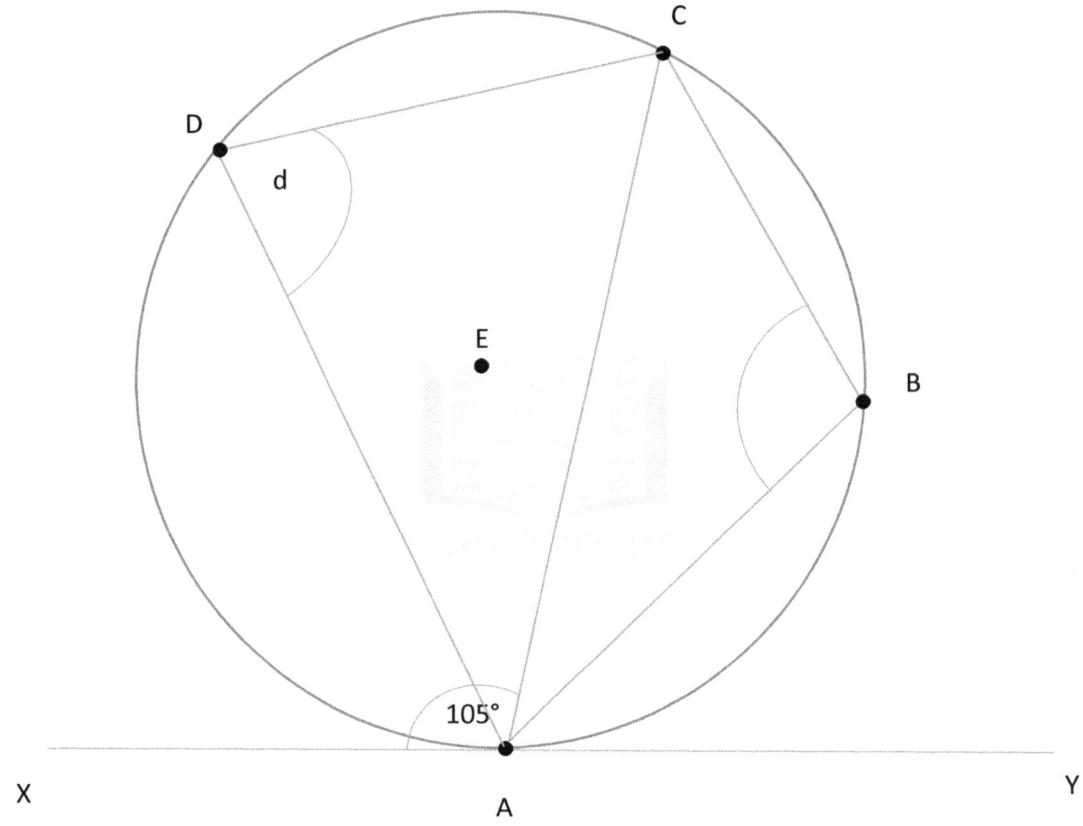

Determine the value of d and give reasons for your answer.

<ABC = 105° (Reason: Alternate Segment Theorem)

d = 180° - 105° **d = 75°** (Reason: Opposite angles in a cyclic quadrilateral sum to 180°.)

(4 marks)

END OF WORKSHEET

868 TUTORS

Preparation for

High School Mathematics

Computation

Solutions

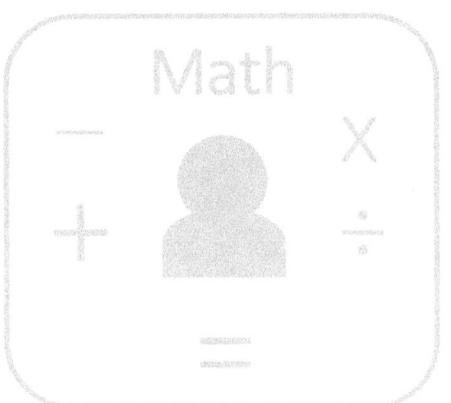

Instructions and Tips:

- ✓ You have 90 minutes to complete this worksheet
- ✓ This worksheet consists of 16 questions
- ✓ Write answers in the spaces provided
- ✓ All working must be clearly shown

Student Name: _____

Student ID: _____

Date: __ / __ / ____

Total Score:

Highest Score:

Tutor's Comments:

Access more free worksheets at www.868tutors.com

Question 1

Using a calculator, calculate the EXACT value of :

(a) $(2¾)^2 \div 2¾$, expressing your answer as a fraction.

$\left(\frac{11}{4}\right)^2 \div \left(\frac{11}{4}\right)$

$\left(\frac{11}{4}\right) \times \left(\frac{11}{4}\right) \div \left(\frac{11}{4}\right)$

$\left(\frac{121}{16}\right) \div \left(\frac{11}{4}\right)$

$\left(\frac{121}{16}\right) \times \left(\frac{4}{11}\right) = \boxed{\frac{11}{4}}$

(3 marks)

(b) $\sqrt{3.24} + 0.32$, expressing your answer in standard form.

$1.8 + 0.32 = 2.12 = \boxed{2.12 \times 10^0}$

(3 marks)

Question 2

Write each value below to 2 decimal places:

(a) 5.954 = 5.95

(b) 3.443 = 3.44

(c) 2.657 = 2.66

(d) 3.999 = 4.00

(e) 4.523 = 4.52

(f) 2.119 = 2.12

(g) 0.011 = 0.01

(h) 0.989 = 0.99

(i) 1.343 = 1.34

(j) 4.112 = 4.11

(5 marks)

Question 3

Write each value below to 2 significant figures:

(a) 5.954 = $\boxed{6.0}$

(b) 3.443 = $\boxed{3.4}$

(c) 2.657 = $\boxed{2.7}$

(d) 3.999 = $\boxed{4.0}$

(e) 4.523 = $\boxed{4.5}$

(f) 2.119 = $\boxed{2.1}$

(g) 0.011 = $\boxed{0.011}$

(h) 0.989 = $\boxed{0.99}$

(i) 1.343 = $\boxed{1.3}$

(j) 4.112 = $\boxed{4.1}$

(5 marks)

Question 4

Write each value below in Standard form (Scientific Notation):

(a) 5974 = 5.974×10^3

(b) 34.46 = 3.446×10^1

(c) 0.0013 = 1.3×10^{-3}

(d) 3.9 = 3.9×10^0

(e) 452.3 = 4.523×10^2

(f) 21.19 = 2.119×10^1

(g) 18 = 1.8×10^1

(h) 0.989 = 9.89×10^{-1}

(i) 13.3 = 1.33×10^1

(j) 0.000002 = 2.0×10^{-6}

(10 marks)

Question 5

(a) Shade $\frac{1}{4}$ of the grid shown below:

(2 marks)

(b) Shade $\frac{1}{5}$ of the grid shown below:

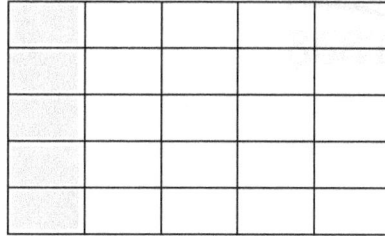

(2 marks)

(c) Shade $\frac{1}{3}$ of the grid shown below:

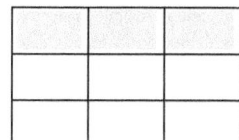

(2 marks)

Question 6

Determine and insert the appropriate numerator so that the following mathematical statements become valid.

(a) $20\% = \dfrac{1}{5}$

(b) $30\% = \dfrac{30}{100}$

(c) $20\% = \dfrac{40}{200}$

(d) $19\% = \dfrac{19}{100}$

(e) $16\% = \dfrac{64}{400}$

(5 marks)

Question 7

Solve the following:

(a) $\dfrac{4}{5}$ of 80 $\dfrac{4}{5} \times \dfrac{80}{1} = \dfrac{4}{1} \times \dfrac{16}{1} = \boxed{64}$

(b) $\dfrac{2}{8}$ of 100 $\dfrac{2}{8} \times \dfrac{100}{1} = \dfrac{1}{4} \times \dfrac{100}{1} = \boxed{25}$

(c) $\dfrac{1}{4}$ of 72 $\dfrac{1}{4} \times \dfrac{72}{1} = \dfrac{1}{1} \times \dfrac{18}{1} = \boxed{18}$

(d) $\dfrac{1}{10}$ of 1000 $\dfrac{1}{1} \times \dfrac{100}{1} = \boxed{100}$

(e) $\dfrac{1}{7}$ of 49 $\dfrac{1}{1} \times \dfrac{7}{1} = \boxed{7}$

(5 marks)

Question 8

(a) A Mathematics textbook is listed at 120 TTD without Value Added Tax (VAT). VAT is introduced on the textbook at 12.5%. Calculate the amount of money that will be spent on VAT.

List price = 120 TTD

$\text{VAT} = \dfrac{12.5}{100} \times 120 \text{ TTD}$

Amount of money spent on VAT = 15 TTD

(3 marks)

(b) A lady saves 20% of her monthly take-home salary. She saves $2500 TTD per month. Calculate her monthly take-home salary.

20 % = 2500 TTD

$1\% = \dfrac{2500 \text{ TTD}}{20}$

$100\% = \dfrac{2500 \text{ TTD}}{20} \times 100$

100 % = 12,500 TTD

Her monthly take home salary = 12,500 TTD

(3 marks)

Question 9

An alloy consists of copper, aluminum, zinc in the ratios 8: 1: 1. Find the mass of each metal in 100 kilograms of the alloy.

Copper: Aluminum: Zinc = 8:1:1

Copper = $\frac{8}{10}$ of 100 kg = $\frac{8}{1} \times 10$ = 80 kg

Aluminum = $\frac{1}{10}$ of 100 kg = $\frac{1}{1} \times 10$ = 10 kg

Zinc = $\frac{1}{10}$ of 100 kg = $\frac{1}{1} \times 10$ = 10 kg

| Mass of Copper = 80 kg |

| Mass of Aluminum = 10 kg |

| Mass of Zinc = 10 kg |

(4 marks)

Question 10

Calculate the exact value of:

$$\frac{2\frac{1}{7} - \frac{5}{21}}{\frac{2}{7} \times \frac{40}{8}}$$

Simplifying the numerator: $\frac{15}{7} - \frac{5}{21}$ → $\frac{45-5}{21} = \frac{40}{21}$

Simplifying the denominator: $\frac{2}{7} \times \frac{40}{8} = \frac{10}{7}$

$$\frac{\frac{40}{21}}{\frac{10}{7}} = \frac{40}{21} \div \frac{10}{7} = \frac{40}{21} \times \frac{7}{10} = \frac{8}{6} = \boxed{\frac{4}{3}}$$

(3 marks)

Question 11

(a) Using a calculator, evaluate:

$6.24(9 - 1.67)$

$\boxed{6.24(7.33) = 45.7392}$

(2 marks)

(b) $\dfrac{10.68}{3.5^2 - 1.45}$

Simplifying the denominator

$3.5^2 - 1.45 = 10.8$

$\dfrac{10.68}{10.8} = 0.988 = \boxed{0.989 \text{ (to 3 decimal places)}}$

(2 marks)

(c) $4 - \dfrac{0.48}{0.15}$

$4 - 3.2 = \boxed{0.8}$

(2 marks)

Question 12

(a) A sum of money is shared between Selina and Sierra in the ratio 3: 2. Selina received $120. How much money was shared altogether?

Selina: Sierra   3:2

Selina = $\frac{3}{5}$

Sierra = $\frac{2}{5}$

Selina received $120.00

Therefore $\frac{3}{5}$ = 120

$\frac{1}{5} = \frac{120}{3}$

$\frac{5}{5} = \frac{120}{3} \times \frac{5}{1}$ = $200   **Total money shared = $200**

(3 marks)

(b) In the Republic of Trinidad and Tobago, 3 litres of diesel cost 5.16 TTD

Calculate the cost of 5 litres of diesel in Trinidad and Tobago.

Cost of 3 litres of diesel = 5.16 TTD

Cost of 1 litre of diesel = $\frac{5.16 \text{ TTD}}{3}$ = 1.72 TTD

Cost of 5 litres of diesel = $\frac{5.16 \text{ TTD}}{5} \times \frac{5}{1}$ = **8.60 TTD**

(2 marks)

(b) How many litres of diesel can be bought for 100.00 TTD in Trinidad and Tobago?

For 5.16 TTD, 3 litres of diesel can be purchased

5.16 TTD = 3 litres

1 TTD = $\frac{3 \text{ litres}}{5.16}$

100 TTD = $\frac{3 \text{ litres}}{5.16} \times \frac{100}{1}$ = **58.14 litres (to 2 decimal places)**

(2 marks)

Question 13

(a) Calculate the EXACT value of:

$4\frac{1}{5} - (1\frac{1}{9} \times 18)$

Give priority to terms inside the brackets

$\frac{10}{9} \times \frac{18}{1} = \frac{20}{1}$

$4\frac{1}{5} - 20$

$\frac{21}{5} - \frac{20}{1}$

$\frac{21-100}{5} = \boxed{\frac{-79}{5}}$

(2 marks)

Question 14

a) **List the first five natural numbers**

1,2,3,4,5

(2 marks)

b) **What does this symbol represent in Mathematics** : $\mathbb{Z}$

This symbol is used to represent Integers in Mathematics

(2 marks)

Question 15

Imru decides to make a grapefruit juice drink. His dad informs him that for every 5 cups of pure grapefruit juice, he should add 4 cups of water and one cup of sugar.

(a) Express this instruction as a ratio (pure juice:water:sugar).

5:4:1

(1 mark)

(b) Imru wants to make 20 cups of grapefruit juice drink. Calculate how many cups of water Imru needs to use to adhere to his father's recipe.

Pure juice = $\frac{5}{10}$

Water = $\frac{4}{10}$

Sugar = $\frac{1}{10}$

$\frac{4}{10} \times 20 = 8$

Imru should use 8 cups of water

(2 marks)

Question 16

(a) An alloy consists of iron and aluminum the ratios 1: 9. Find the mass of each metal in 50 kilograms of the alloy.

Iron: Aluminum 1:9

Iron = $\frac{1}{10}$

Aluminum = $\frac{9}{10}$

Mass of iron = $\frac{1}{10} \times \frac{50 \text{ kg}}{1}$ = **5 kg**

Mass of aluminum = $\frac{9}{10} \times \frac{50 \text{ kg}}{1}$ = **45 kg**

(2 marks)

(b) An alloy consists of copper and zinc in the ratio 4:1. Find the amount of each metal in 40 kilograms of the alloy.

Copper: Zinc 4:1

Copper = $\frac{4}{5}$ of 40 kg = $\frac{4}{5} \times 40$ kg = 32 kg

Zinc = $\frac{1}{5}$ of 40 kg = $\frac{1}{5} \times 40$ kg = 8 kg

Mass of copper = **32 kg**

Mass of zinc = $\frac{9}{10} \times \frac{50 \text{ kg}}{1}$ = **8 kg**

(2 marks)

END OF WORKSHEET

868

868TUTORS

Preparation for

High School Mathematics

Consumer Arithmetic

Solutions

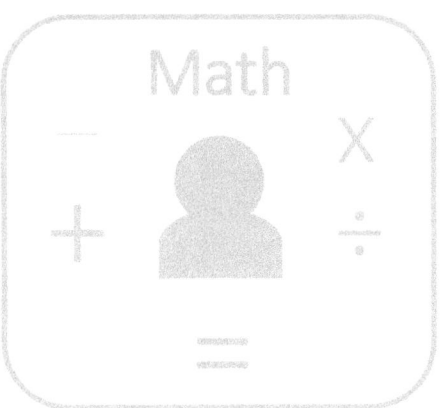

Instructions and Tips:

- ✓ You have 90 minutes to complete this worksheet
- ✓ This worksheet consists of 8 questions
- ✓ Write answers in the spaces provided
- ✓ All working must be clearly shown
- ✓ Answers should be given to 2 decimal places

Student Name: _____

Student ID: _____

Date: __/__/____

Total Score:

Highest Score:

Tutor's Comments:

Purchase solutions at www.868tutors.com

Question 1

Chukwuemeka arrives in Trinidad and Tobago on vacation. He has 50,000 Nigerian Naira (NGN). He changes the 50,000 NGN at a local bank to Trinidad and Tobago Dollars (TTD).

1 TTD = 30.20 NGN

(a) How much TTD should Chukwuemeka receive?

30.20 NGN $= 1$ TTD

1 NGN $= \frac{1 \text{ TTD}}{30.20}$

$50,000$ NGN $= \frac{1 \text{ TTD}}{30.20} \times 50,000$

$50,000$ NGN $= 1,655.629$ TTD

Chukwuemeka should receive 1,655.63 TTD

(2 marks)

During his stay in Trinidad and Tobago, Chukwuemeka spends 600 TTD. He then travels to Barbados. He converts the remainder of his TTD to the Barbados Dollar (BDS).

1 BDS = 3.22 TTD

(b) How much BDS should Chukwuemeka receive?

Amount of money remaining = Original sum - Money spent

Amount of money remaining = 1,655.63 TTD - 600 TTD

Amount of money remaining = 1,055.63 TTD

(Converting to BDS) 3.22 TTD $= 1$ BDS

1 TTD $= \frac{1 \text{ BDS}}{3.22}$

$1,055.63$ TTD $= \frac{1 \text{ BDS}}{3.22} \times 1,055.63 = 327.84$ BDS

Chukwuemeka should receive 327.84 BDS

(2 marks)

Question 2

Option A: Cost 50 TTD

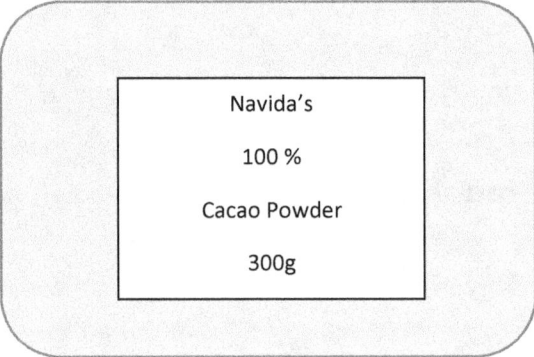

Option B: Cost 100 TTD

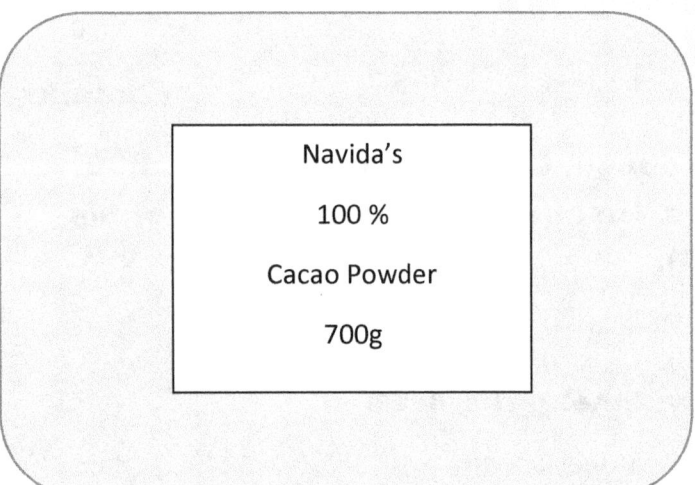

At a local grocery store you observe that there are two sizes of cacao powder packs. Both are of the identical brand and identical quality. Calculate the cost per gram for:

(a) Option A

$300 \text{ g} = 50 \text{ TTD}$

$1 \text{ g} = \frac{50 \text{ TTD}}{300}$

$1 \text{ g} = 0.166 \text{ TTD}$

$\boxed{1 \text{ g} = 0.17 \text{ TTD}}$

(2 marks)

(b) Option B

$700 \text{ g} = 100 \text{ TTD}$

$1 \text{ g} = \frac{100 \text{ TTD}}{700}$

$1 \text{ g} = 0.1428 \text{ TTD}$

$\boxed{1 \text{ g} = 0.14 \text{ TTD}}$

(2 marks)

(c) Based on your calculations, recommend which option is more cost effective given that you purchase cacao powder regularly and that you have the cash to purchase either option.

Option B is more cost effective since the cost per gram with Option B (0.14 TTD) is lower than the cost per gram with Option A (0.17 TTD).

(1 mark)

Question 3

At a supermarket in Belize, a 2 Litre bottle of Coconut water can be purchased for 1.50 Belize Dollars (BZD).

1 BZD = 1.35 XCD

(a) Calculate how much Eastern Caribbean Dollars (XCD) will be needed to purchase this Coconut water in Belize.

1 BZD = 1.35 XCD

1.50 BZD = 1.35 XCD × 1.50

1.50 BZD = 2.025 XCD

2.03 XCD will be needed to purchase this Coconut water in Belize.

(1 mark)

1 BZD = 0.50 USD

(b) Calculate the cost per litre of Coconut water in USD.

Cost for 2 litres of coconut water = 1.50 BZD

Cost for 1 litre of coconut water = $\frac{1.50 \text{ BZD}}{2}$ = 0.75 BZD

Converting BZD to USD

1 BZD = 0.50 USD

0.75 BZD = 0.50 USD × 0.75 = 0.375 USD

Cost per litre of coconut water = 0.38 USD

(2 marks)

1 USD = 6.51 TTD

(c) A businessman in Trinidad and Tobago plans to sell the 2 litre bottle of Coconut water with a 30% markup. Calculate the cost in TTD for a bottle of this Coconut water with the included markup.

Cost of Coconut Water = 1.50 BZD	*Cost of coconut water = 4.88 TTD*
1 BZD = 0.50 USD	*30 % markup = $\frac{30}{100}$ × 4.88 = 1.46 TTD*
1.50 BZD = 0.50 USD × 1.50 = 0.75 USD	*Total cost = 4.88 TTD + 1.46 TTD*
1 USD = 6.51 TTD	**Total = 6.34 TTD**
0.75 USD = 4.88 TTD	

(3 marks)

(d) At the point of sale in Trinidad and Tobago, Value Added Tax (VAT) of 12.5% will be added. Calculate the cost in TTD for a 2 litre bottle of this Coconut water, inclusive of the markup and VAT.

Cost inclusive of markup = 6.34 TTD

VAT (12.5 %) = $\frac{12.5}{100}$ × 6.34 TTD = 0.79 TTD

Total Cost = Cost inclusive of markup + VAT

Total Cost = 6.34 TTD + 0.79 TTD = 7.13 TTD

(4 marks)

Question 4

A sports store in Jamaica sells a bicycle for 18,000 Jamaican Dollars (JMD)

(a) At Christmas time, the customer is given a 20% discount. Calculate the cost of the bicycle at this time.

New Price = Original Price – Discount

New Price = 18,000 JMD – 3,600 JMD

New Price = 14,400 JMD

(2 marks)

(b) For every bicycle, the store sells during Christmas time, a profit of 4,500 JMD is made. If the store sells 150 bicycles, what is the store's overall profit from this bicycle?

Profit from 1 bicycle = 4,500 JMD

Profit from 150 bicycles = 4,500 JMD × 150 = 675,000 JMD

Overall profit from this bicycle = 675,000 JMD

(2 marks)

1 TTD = 18.77 JMD

(c) Calculate the stores' profit from this bicycle in TTD.

18.77 JMD = 1 TTD

$1 \text{ JMD} = \frac{1 \text{ TTD}}{18.77}$

$675,000 \text{ JMD} = \frac{1 \text{ TTD}}{18.77} \times 675,000$

675,000 JMD = **35,961.64 TTD**

(2 marks)

Question 5

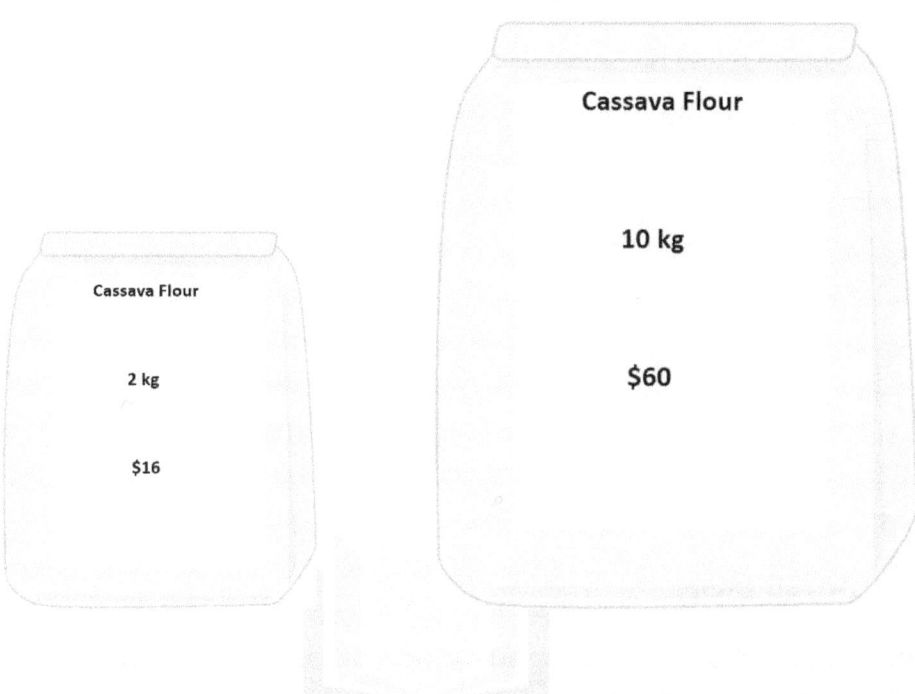

Mr. Kalesanwo decides to open a Nigerian restaurant in Trinidad. At a grocery store, he notices two options for purchasing Cassava Flour. Both options are of the same brand and quality. The prices are quoted in Trinidad and Tobago Dollars.

(a) Calculate the cost of purchasing 30 kg of Cassava Flour using 2 kg packs (in TTD).

Required mass of flour = 30 kg

Number of packs required (2 kg) = $\frac{30 \text{ kg}}{2 \text{ kg}}$ = 15 packs

Cost of one 2 kg pack = 16 TTD

Cost of fifteen 2 kg packs = 16 TTD × 15

Cost of fifteen 2 kg packs = 240 TTD

(2 marks)

(b) Calculate the cost of purchasing 30 kg of Cassava Flour using 10 kg packs.

Required mass of flour = 30 kg

Number of packs required (10 kg) = $\frac{30 \text{ kg}}{10 \text{ kg}}$ = 3 packs

Cost of one 10 kg pack = 60 TTD

Cost of three 10 kg packs = 60 TTD × 3

Cost of three 10 kg packs = 180 TTD

(2 marks)

1 TTD = 30.20 NGN

(c) Calculate the cost of the cheaper option for 30 kg of Cassava Flour in Nigerian Naira (NGN).

It is cheaper to purchase 30 kg of Cassava Flour by using the 10 kg packs.

Converting 180 TTD to NGN

1 TTD = 30.20 NGN

180 TTD = 30.20 NGN × 180

180 TTD = **5,436 NGN**

(2 marks)

Question 6

A young Electrical Engineer is hired in the oil and gas industry in Trinidad and Tobago. Her monthly salary consists of a base pay of 15 000 TTD. In addition, every time she is called offshore, she will be paid an additional 2000 TTD.

(a) Calculate her annual salary for that year if she is called offshore 15 times for the year (in TTD).

Base pay for 1 month = 15,000 TTD

Base pay for 12 months = 15,000 TTD × 12 = 180,000 TTD

Earning from one offshore call = 2,000 TTD

Earning from fifteen offshore calls = 2,000 TTD × 15 = 30,000 TTD

Salary for year = 180,000 TTD + 30,000 TTD = 210,000 TTD

(2 marks)

(b) As a citizen of Trinidad and Tobago, the young engineer is entitled to an annual personal allowance of 72,000 TTD. Calculate the amount of her salary that she will be taxed on for that year (in TTD).

Amount that will be taxed = Annual Salary − personal allowance

= 210,000 TTD − 72,000 TTD

Amount that will be taxed = 138,000 TTD

(2 marks)

(c) The engineer is taxed at a rate of 25%. Calculate, the amount of money that she takes home for that year (in TTD), after being taxed.

Amount that will be taxed = 138,000 TTD

Tax amount = $\frac{25}{100}$ × 138,000 TTD = 34,500 TTD

Annual take home = Annual Salary − Tax

Annual take home = 210,000 TTD − 34,500 TTD

Annual take home = 175,500 TTD

(2 marks)

Question 7

Mr and Mrs. Yang visit Tobago and are awestruck with the beauty of the island. They decide to purchase an acre of beach front land. The cost is 500 TTD per square foot.

1 Acre = 43,560 square feet.

(a) Calculate the cost in TTD of one acre of the beach front land.

Cost of 1 square foot = 500 TTD

Cost of 43,560 square feet = 500 TTD × 43,560

Cost of 43,560 square feet = 21,780,000.00 TTD

Cost of one acre = 21,780,000.00 TTD

(2 marks)

(b) Stamp duty will be charged at a rate of 2% of the cost of the land. Calculate the amount that Mr. and Mrs. Yang will have to pay in Stamp duty.

Cost of land = 21,780,000.00 TTD

Stamp Duty = $\frac{2}{100}$ × 21,780,000.00 TTD

Stamp Duty = 435,600 TTD

(2 marks)

1 USD = 6.51 TTD

(c) Calculate the total amount of money that Mr. and Mrs. Yang will spend in USD.

Total amount = Cost of one acre + Stamp duty

= 21,780,000 TTD + 435,600 TTD

Total amount = 22,215,600 TTD

6.51 TTD = 1 USD

1 TTD = $\frac{1 \text{ USD}}{6.51}$

22,215,600 TTD = $\frac{1 \text{ USD}}{6.51}$ × 22,215,600 = **3,412,534.56 USD**

(2 marks)

Question 8

(a) The table below shows Adrian's shopping bill for his convenience store. Some information was not included.

Items	Quantity	Unit Price (TTD)	Total Cost (TTD)
Flash drives	15	20.00	A
Mobile phones	40	B	12,000
Cranberry flavored Water	C litres	4.35	435
Sub-Total			12,735
12.5% VAT (to the nearest cent)			D

Calculate the values of A, B, C, and D

A: Price of 1 Flash drive = 20.00 TTD

Price of 15 Flash drives = 20.00 TTD × 15

Price of 15 Flash drives = $\boxed{300.00 \text{ TTD}}$

B: Total cost (40 mobile phones) = 12,000 TTD

Cost of 1 mobile phone (Unit price) = $\frac{12,000}{40}$

= $\boxed{300 \text{ TTD}}$

C: Total cost = 435 TTD

Unit cost (litre) = 4.35 TTD

Number of litres = $\frac{435 \text{ TTD}}{4.35 \text{ TTD}}$

= $\boxed{100}$

D: 12.5% VAT

$\frac{12.5}{100} \times 12{,}735 = \boxed{1591.88 \text{ TTD}}$

(5 marks)

(b) VAT was increased from 12.5% to 20%. Calculate the increase in Adrian's bill.

12.5% VAT

$$\frac{12.5}{100} \times 12{,}735 = 1591.88 \text{ TTD}$$

20 % VAT

$$\frac{20}{100} \times 12{,}735 = 2{,}547 \text{ TTD}$$

Increase in Adrian's bill = 2,547 TTD − 1591.88 TTD
Increase in Adrian's bill = **955.12 TTD**

(2 marks)

END OF WORKSHEET

868

868TUTORS

TUTORS

Preparation for

High School Mathematics

Consumer Arithmetic II

Solutions

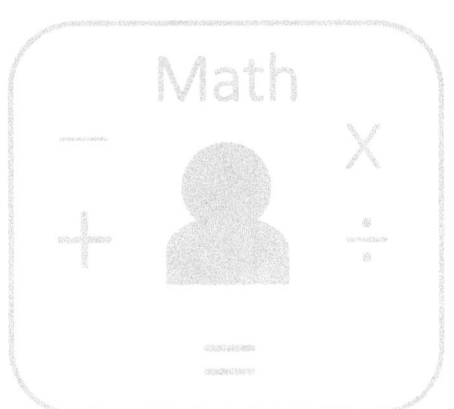

Instructions and Tips:

- ✓ You have 60 minutes to complete this worksheet
- ✓ This worksheet consists of 7 questions
- ✓ Write answers in the spaces provided
- ✓ All working must be clearly shown
- ✓ Answers should be given to 2 decimal places

Student Name: _____

Student ID: _____

Date: __/__/____

Total Score:

Highest Score:

Tutor's Comments:

Access more free worksheets at www.868tutors.com

Question 1

(a) The table below shows Georgia's shopping bill. Some information was not included

Items	Quantity	Unit Price (TTD)	Total Cost (TTD)
Cassava flour	6.5 kg	2.40	A
Pimentos	6 bags	B	52.80
Coconut Water	C liters	12.35	98.80
Sub-Total			167.20
15% VAT (to the nearest cent)			D

Calculate the values of A, B, C, and D

A: *Unit price of Cassava Flour (1 kg) = 2.40 TTD*

$\quad$ *6.5 kg = 2.40 TTD × 6.5*
$\quad$ *6.5 kg =* **15.60 TTD**

B: *Total cost (6 bags) = 52.80 TTD*

Cost of 1 bag = $\frac{52.80}{6}$ = **8.80 TTD**

C: *Total cost = 98.80*

$\quad$ *Unit cost (liters) = 12.35 TTD*
$\quad$ *Number of litres =* $\frac{98.80}{12.35 \, TTD}$

$\quad$ *Number of litres =* **8**

D: *15% VAT*

$\quad$ $\frac{15}{100}$ × 167.20 = **25.08 TTD**

(5 marks)

(b) VAT was reduced from 15% to 12.5%. Calculate the reduction in Georgia's bill.

15% VAT

$\frac{15}{100} \times 167.20 =$ **25.08 TTD**

12.5% VAT

$\frac{12.5}{100} \times 167.20 =$ **20.90 TTD**

Reduction in Georgia's bill = 25.08 TTD - 20.90 TTD

Reduction in Georgia's bill = **4.18 TTD**

(2 marks)

Question 2

(a) How much simple interest is due on a loan of $14,000 for two years if the annual rate of interest is ½ percent?

$$Simple\ Interest = \frac{Principal \times Rate \times Time}{100}$$

$$Simple\ Interest = \frac{14{,}000 \times 0.5 \times 2}{100}$$

Simple Interest = $140

Principal = 14,000
Rate = 0.5%
Number of years = 2

(2 marks)

(b) How much simple interest is due on a loan of $20,000 for three years if the annual rate of interest is 5 percent?

$$Simple\ Interest = \frac{Principal \times Rate \times Time}{100}$$

$$Simple\ Interest = \frac{20{,}000 \times 5 \times 3}{100}$$

Simple Interest = $3000

Principal = 20,000
Rate = 5%
Number of years = 3

(2 marks)

Question 3

(a) In the Republic of Trinidad and Tobago, 3 litres of diesel cost TT$5.16

Calculate the cost of 5 litres of diesel in Trinidad and Tobago

Cost of 3 litres of diesel = 5.16 TTD

Cost of 1 litre of diesel = $\frac{5.16 \text{ TTD}}{3}$

Cost of 5 litres of diesel = $\frac{5.16 \text{ TTD}}{3} \times 5$

Cost of 5 litres of diesel = 8.60 TTD

(2 marks)

(b) How many litres of diesel can be bought for TT$100.00 in Trinidad and Tobago?

5.16 TTD = 3 litres of diesel

1 TTD = $\frac{3 \text{ litres}}{5.16}$

100 TTD = $\frac{3 \text{ litres}}{5.16} \times 100$

100 TTD = 58.14 litres

(2 marks)

Question 4

A man in Barbados invests 12,000 Barbados Dollars (BDS) into an account that pays 8.5% interest per year, compounded annually. Calculate the amount of money that he will have after 3 years.

Compound Interest Formula

Amount = Principal $\left(1 + \frac{\text{Rate}}{100}\right)^{\text{number of years}}$

Amount = $12{,}000 \left(1 + \frac{8.5}{100}\right)^n$

Principal = 12,000
Rate = 8.5%
Number of years = 3

Amount = $12{,}000 \, (1.085)^3 = 15{,}327.47$

Amount of money after 3 years = 15,327.47 BDS

(3 marks)

Question 5

A man in Guyana invests 15,000 Guyanese Dollars (GYD) into an account that pays 9.5% interest per year, compounded annually. Calculate the amount of money that he will have after 2 years.

Compound Interest Formula

$Amount = Principal \left(1 + \frac{Rate}{100}\right)^{number\ of\ years}$

$Amount = 15,000 \left(1 + \frac{9.5}{100}\right)^2$

Principal = 15,000
Rate = 9.5%
Number of years = 2

$Amount = 15,000 (1.095)^2 = 17,985.38$

Amount of money after 2 years = 17,985.38 GYD

(3 marks)

Question 6

A farmer purchases a pickup truck for $280,000. The pickup truck depreciates at a rate of 5% per year. Determine the value of the pickup truck after 4 years?

Depreciation formula

$Value = Purchase\ Price \left(1 - \frac{Rate}{100}\right)^{number\ of\ years}$

$Value = 280{,}000 \left(1 - \frac{5}{100}\right)^4$

$Value = 280{,}000\ (0.95)^4$

$Value = 280{,}000\ (0.81450625)$

$Value = \$228{,}061.75$

Value of pickup truck after 4 years = $228,061.75

Purchase Price = 280,000

Rate = 5%

Number of years = 4

(4 marks)

Question 7

The interest rate on savings in a bank decreased from 5 ½ percent per annum to 4 percent per annum. Calculate the difference in annual interest on a deposit of $10,000.

Calculating Simple Interest at 5 ½ %

Simple Interest = $\frac{Principal \times Rate \times Time}{100}$

Simple Interest = $\frac{10,000 \times 5.5 \times 1}{100}$ = $550

Calculating Simple Interest at 4 %

Simple Interest = $\frac{Principal \times Rate \times Time}{100}$

Simple Interest = $\frac{10,000 \times 4 \times 1}{100}$ = $400

Difference in annual interest on deposit = $550 - $400 = **$150**

(4 marks)

END OF WORKSHEET

868 TUTORS

Preparation for

High School Mathematics

Functions

Solutions

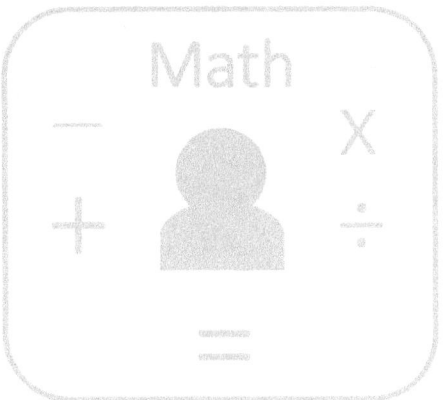

Instructions and Tips:

- ✓ You have 60 minutes to complete this worksheet
- ✓ This worksheet consists of 6 questions
- ✓ Write answers in the spaces provided
- ✓ Show all working

Student Name: _____

Student ID: _____

Date: __ / __ / ____

Total Score:

Highest Score:

Tutor's Comments:

Access more free worksheets at www.868tutors.com

Question 1

Given that f(x) = x + 1 and g(x) = $\frac{2}{x}$

(a) Calculate f(2).

f(x) = x + 1

f(2) = 2 + 1

$\boxed{f(2) = 3}$

(b) Calculate g(2).

g(x) = $\frac{2}{x}$ g(2) = $\frac{2}{2}$

$\boxed{g(2) = 1}$

(c) Calculate g(-1).

g(x) = $\frac{2}{x}$ g(-1) = $\frac{2}{-1}$

$\boxed{g(-1) = -2}$

(d) Write an expression for gf(x).

$\boxed{gf(x) = \frac{2}{x+1}}$

(e) Write an expression for fg(x).

fg(x) = $\frac{2}{x} + \frac{1}{1}$

$\boxed{fg(x) = \frac{2+x}{x}}$

(f) Calculate fg(2).

$f(x) = x + 1$

$g(x) = \frac{2}{x}$

$fg(x) = \frac{2 + x}{x}$ $fg(2) = \frac{2 + 2}{2}$

$\boxed{fg(2) = 2}$

(g) Calculate the values of x so that f(x) = g(x).

$f(x) = x + 1$ $g(x) = \frac{2}{x}$

$x + 1 = \frac{2}{x}$

$\frac{x+1}{1} = \frac{2}{x}$ (By cross-multiplication)

$x(x+1) = 2$

$x^2 + x = 2$

$x^2 + x - 2 = 0$

$(x + 2)(x - 1) = 0$

$\boxed{x = -2 \text{ or } x = 1}$

(10 marks)

Question 2

Given that f(x) = x + 3 and g(x) = $\frac{4}{x}$

(a) Calculate f(-2).

f(x) = x + 3

f(-2) = -2 + 3

$\boxed{f(-2) = 1}$

(b) Calculate g(2).

g(x) = $\frac{4}{x}$ g(2) = $\frac{4}{2}$

$\boxed{g(2) = 2}$

(c) Calculate g(-2).

g(-2) = $\frac{4}{-2}$

$\boxed{g(-2) = -2}$

(d) Write an expression for gf(x).

$\boxed{gf(x) = \frac{4}{x+3}}$

(e) Write an expression for fg(x).

fg(x) = $\frac{4}{x} + \frac{3}{1}$

$\boxed{fg(x) = \frac{4+3x}{x}}$

(f) Calculate fg(1).

$$fg(x) = \frac{4+3x}{x} \quad fg(1) = \frac{4+3(1)}{1}$$

$\boxed{fg(1) = 7}$

(g) Calculate the values of x so that f(x) = g(x).

$f(x) = x + 3 \quad g(x) = \frac{4}{x}$

$x + 3 = \frac{4}{x}$

$\frac{x+3}{1} = \frac{4}{x}$ (By cross-multiplication)

$x(x+3) = 4$

$x^2 + 3x = 4$

$x^2 + 3x - 4 = 0$

$(x+4)(x-1) = 0$

$\boxed{x = -4 \text{ or } x = 1}$

(10 marks)

Question 3

Given that f(x) = x^2 and g(x) = x - 3

(a) Calculate g(6).

g(x) = x - 3

g(6) = 6 - 3

$\boxed{g(6) = 3}$

(b) Calculate $g^{-1}(4)$.

g(x) = x - 3

let y = x - 3

x - 3 = y   (replace x with y and replace y with x)

y - 3 = x

y = x + 3   (make y the subject of the formula)

$g^{-1}(x) = x + 3$

$g^{-1}(4) = 4 + 3$

$\boxed{g^{-1}(4) = 7}$

(c) Write an expression for fg(x).

fg(x) = $(x-3)^2$

fg(x) = (x-3)(x-3)

fg(x) = x^2 -3x -3x + 9

$\boxed{fg(x) = x^2 - 6x + 9}$

(d) Calculate fg(0).

fg(x) = x^2 - 6x + 9

fg(0) = $(0)^2$ - 6(0) + 9

$\boxed{fg(0) = 9}$

(6 marks)

Question 4

Given that $f(x) = \frac{x}{2} - 1$ and $g(x) = 3x + 1$

(a) Calculate g(3).

$g(x) = 3x + 1$

$g(3) = 3(3) + 1$

$\boxed{g(3) = 10}$

Express in its simplest form

(b) $f^{-1}(x)$

$f(x) = \frac{x}{2} - 1$

let $y = \frac{x}{2} - 1$

$x = \frac{y}{2} - 1$ (replace x with y and replace y with x)

$\frac{y}{2} - 1 = x$ (make y the subject of the formula)

$\frac{y}{2} = x + 1$

$\frac{y}{2} = \frac{x + 1}{1}$

$y = 2(x + 1)$

$\boxed{f^{-1}(x) = 2(x + 1)}$

(c) $g^{-1}(x)$

$g(x) = 3x + 1$ let $y = 3x + 1$

$x = 3y + 1$ (replace x with y and replace y with x)

$3y + 1 = x$ $3y = x - 1$ $y = \frac{x - 1}{3}$ (make y the subject of the formula)

$\boxed{g^{-1}(x) = \frac{x - 1}{3}}$

(d) fg(x)

$f(x) = \frac{x}{2} - 1 \quad g(x) = 3x + 1$

$fg(x) = \frac{3x+1}{2} - 1$

$fg(x) = \frac{3x+1}{2} - \frac{2}{2}$

$fg(x) = \frac{3x+1-2}{2}$

$\boxed{fg(x) = \frac{3x-1}{2}}$

(e) (fg)⁻¹ (x)

$fg(x) = \frac{3x-1}{2}$

let $y = \frac{3x-1}{2}$

$x = \frac{3y-1}{2}$ (replace x with y and replace y with x)

$\frac{x}{1} = \frac{3y-1}{2}$ (make y the subject of the formula)

$\frac{x}{1} \bowtie \frac{3y-1}{2}$ (By cross-multiplication)

$3y - 1 = 2x$

$3y = 2x + 1$

$y = \frac{2x+1}{3}$

$\boxed{(fg)^{-1}(x) = \frac{2x+1}{3}}$

(10 marks)

Question 5

Given that f(x) = x + 2 and g(x) = 3x + 4

(a) Calculate f(4).

f(x) = x + 2

f(4) = 4 + 2

$\boxed{f(4) = 6}$

Express in its simplest form

(b) f $^{-1}$ (x)

f(x) = x + 2

let y = x + 2

x = y + 2   (replace x with y and replace y with x)

y = x - 2   (make y the subject of the formula)

$\boxed{f^{-1}(x) = x - 2}$

(c) g^{-1} (x)

g(x) = 3x + 4

let y = 3x + 4

x = 3y + 4   (replace x with y and replace y with x)

3y + 4 = x   (make y the subject of the formula)

3y = x - 4

y = $\frac{x-4}{3}$

$\boxed{g^{-1}(x) = \frac{x-4}{3}}$

(d) fg(x)

$f(x) = x + 2 \quad g(x) = 3x + 4$

$fg(x) = 3x + 4 + 2$

$\boxed{fg(x) = 3x + 6}$

(e) $(fg)^{-1}(x)$

$fg(x) = 3x + 6$

let $y = 3x + 6$

$x = 3y + 6$ (interchange y and x)

$3y = x - 6$ (make y the subject of the formula)

$y = \frac{x-6}{3}$

$\boxed{(fg)^{-1}(x) = \frac{x-6}{3}}$

(f) Show that $(fg)^{-1}(x) = g^{-1}f^{-1}(x)$

Required to show that $(fg)^{-1}(x) = g^{-1}f^{-1}(x)$

$(fg)^{-1}(x) = \frac{x-6}{3}$

recall from (c) $g^{-1}(x) = \frac{x-4}{3}$ recall from (b) $f^{-1}(x) = x - 2$

$g^{-1}f^{-1}(x) = \frac{x-2-4}{3} \quad g^{-1}f^{-1}(x) = \frac{x-6}{3}$

Therefore $(fg)^{-1}(x) = g^{-1}f^{-1}(x)$ has been proven.

(10 marks)

Question 6

Given that g(x) = 2 - x   and f(x) = x³

(a) Calculate f(-3).

f(x) = x³

f(-3) = (-3)³ = -3 × -3 × -3

f(-3) = -27

(b) Calculate fg(3).

fg(x) = (2 - x)³

fg(3) = (2 - 3)³

fg(3) = (2 - 3)³

fg(3) = (-1)³

fg(3) = -1

(c) Calculate gf(3).

gf(x) = 2 - x³

gf(3) = 2 - (3)³

gf(3) = 2 - 27

gf(3) = -25

(6 marks)

END OF WORKSHEET

868

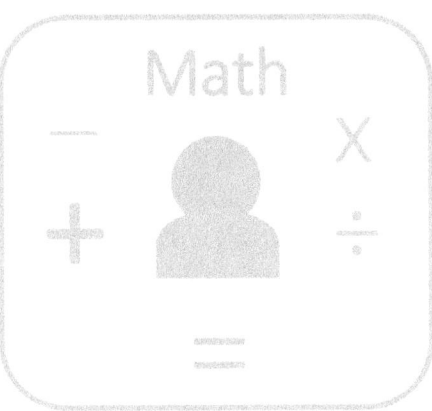

868TUTORS

Preparation for

High School Mathematics

Investigation

Solutions

Instructions and Tips:

- ✓ You have 60 minutes to complete this worksheet
- ✓ This worksheet consists of 6 questions
- ✓ Write answers in the spaces provided
- ✓ Show all working

Student Name: _____

Student ID: _____

Date: __/__/____

Total Score:

Highest Score:

Tutor's Comments:

Access more free worksheets at www.868tutors.com

Question 1

Consider these patterns made from squares:

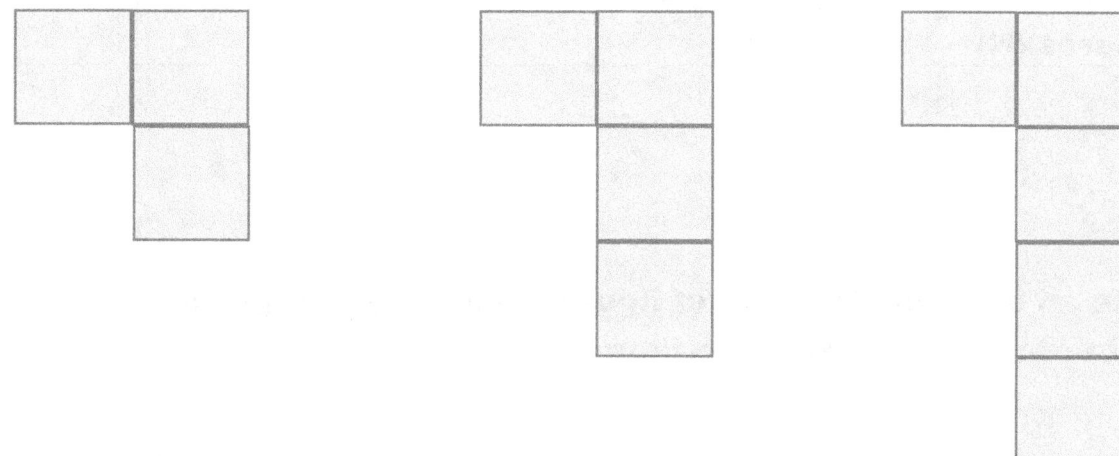

first figure   second figure   third figure

(a) Draw the fourth figure in the pattern :

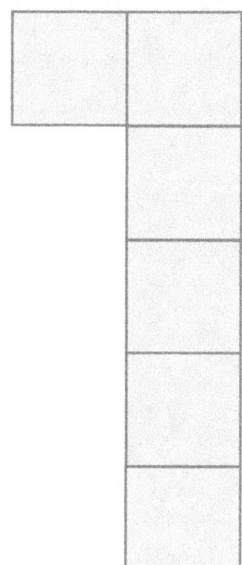

fourth figure

(2 marks)

(b) Complete the table below:

Figure number (f)	1	2	3	4	5	6
Number of squares (N)	3	4	5	6	7	8

(3 marks)

(c) Write an equation in terms of figure number (f) that gives the number of squares (N) in the figure.

$$N = f + 2$$

(3 marks)

(d) Determine the number of squares used for the 12th figure in the pattern.

$N = f + 2$

$N = 12 + 2$

$N = 14$

For the 12th figure, 14 squares are used.

(2 marks)

Question 2

Consider these patterns that consists of triangles and squares:

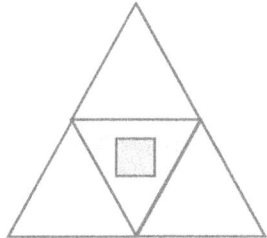

first figure

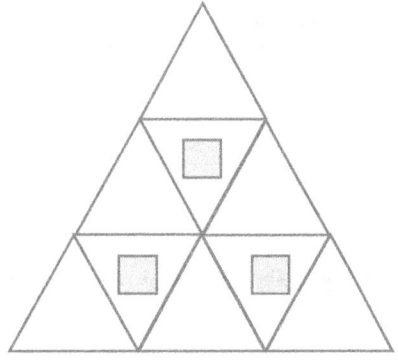

second figure

(a) Draw the third figure in the pattern :

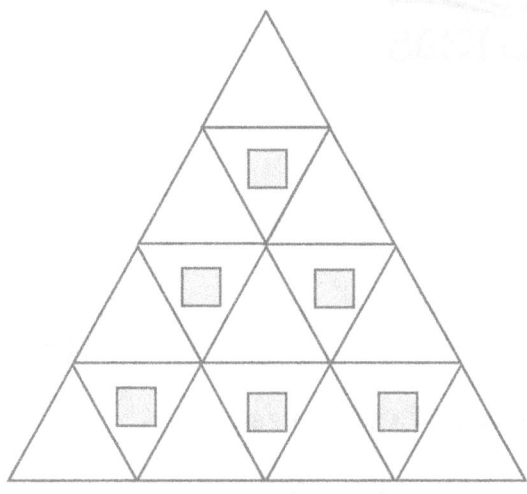

third figure

(3 marks)

(b) Complete the table below :

Figure number (f)	1	2	3
Number of shaded squares (n)	1	3	6

(2 marks)

Consider the sequence of numbers below:

0, 1, 1, 2, 3, 5, 8, 13, 21...

(c) Determine the next three terms in the sequence of numbers.

The next three terms in the sequence of numbers are 34, 55 and 89.

(3 marks)

(d) What is the name of this sequence of numbers?

This sequence of numbers is known as a Fibonacci sequence.

(2 marks)

Question 3

Consider some patterns made from rectangles and triangles:

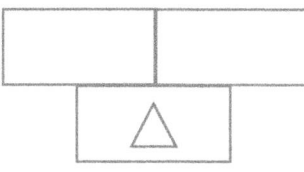   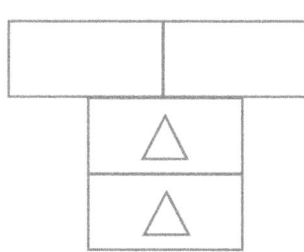   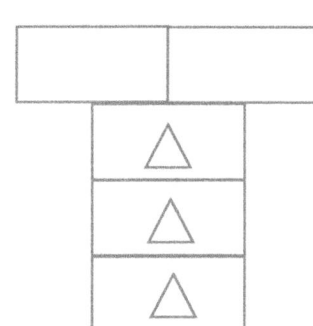

first figure   second figure   third figure

(a) Draw the fourth figure in the pattern :

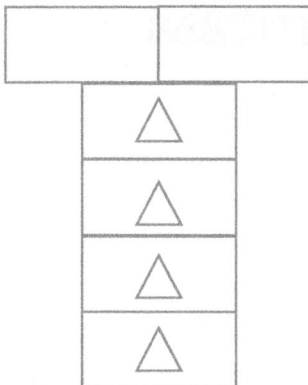

fourth figure

(2 marks)

(b) Draw the fifth figure in the pattern :

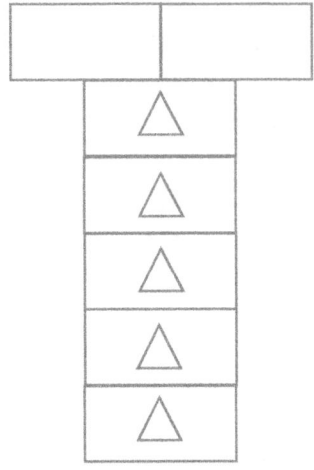

fifth figure

(2 marks)

(c) Complete the table below :

Figure number	1	2	3	4	5	6	7
Number of triangles	1	2	3	4	5	6	7

(2 marks)

(d) How many triangles will the 100th figure have?

The 100th figure will have 100 triangles.

(2 marks)

Question 4

Consider the first three figures in the pattern below:

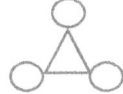   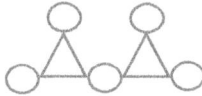   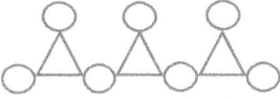

first figure   second figure   third figure

(a) Draw the fourth figure in the pattern :

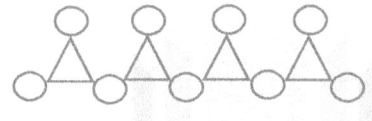

fourth figure

(2 marks)

(b) From your observation and analysis of the patterns, complete the table below:

Figure Number	Number of triangles	Number of circles
1	1	3
2	2	5
3	3	7
4	(i) 4	(ii) 9
n	n	(iii) 2n + 1
(iv) 150	(v) 150	301

$2n + 1 = 301$ $2n = 301 - 1$ $2n = 300$ $n = 150$

(6 marks)

Question 5

Consider the first three figures in the pattern below:

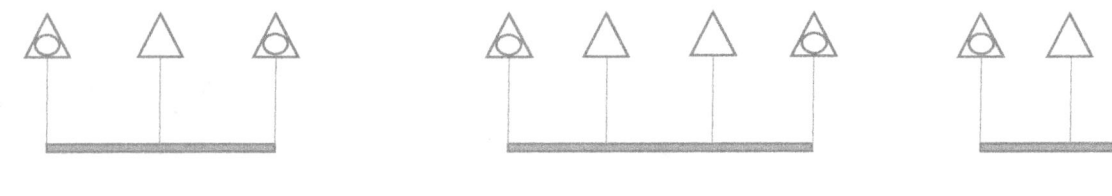

first figure　　　　　　　　second figure　　　　　　　　third figure

(a) Draw the fourth figure in the pattern:

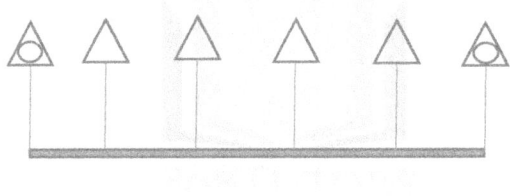

fourth figure

(4 marks)

(b) From your observation and analysis of the patterns, complete the table below:

Figure Number	Number of circles	Number of triangles
1	2	3
2	2	4
3	2	5
4	(i)　2	6
n	2	(ii)　n + 2
(iii)　296	2	298

n + 2 = 298　　n = 298 − 2　　n = 296

(6 marks)

Question 6

first figure second figure third figure

(a) Draw the fourth figure in the pattern :

fourth figure

(3 marks)

(b) Complete the table

Figure Number	Number of horizontal lines	Number of vertical lines
1	1	1
2	2	1
3	2	2
4	(i) ___3___	2
5	3	(ii) ___3___

(2 marks)

END OF WORKSHEET

868

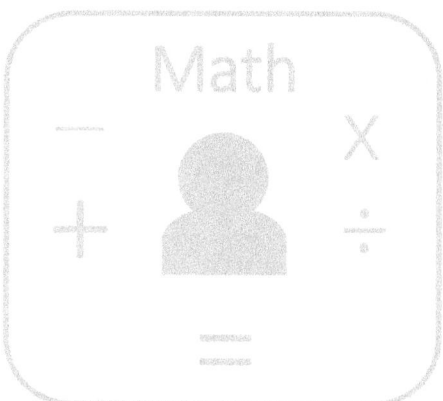

Instructions and Tips:

- ✓ You have 120 minutes to complete this worksheet
- ✓ This worksheet consists of 15 questions
- ✓ Write answers in the spaces provided
- ✓ All working must be clearly shown

868TUTORS

Preparation for

High School Mathematics

Matrices

Solutions

Student Name: _____

Student ID: _____

Date: __ / __ / ____

Total Score:

Highest Score:

Tutor's Comments:

Access more free worksheets at www.868tutors.com

Question 1

State the order of each matrix below:

(a) $\begin{pmatrix} 4 & 2 & 3 \\ 3 & 6 & 6 \\ 1 & 6 & 3 \end{pmatrix}$ number of rows × number of columns

$\boxed{3 \times 3}$

(b) $\begin{pmatrix} 3 \\ 2 \\ 5 \end{pmatrix}$ number of rows × number of columns

$\boxed{3 \times 1}$

(c) $\begin{pmatrix} 2 & 3 & 4 \end{pmatrix}$ number of rows × number of columns

$\boxed{1 \times 3}$

(d) $\begin{pmatrix} 3 \\ 2 \\ 3 \\ 6 \end{pmatrix}$ number of rows × number of columns

$\boxed{4 \times 1}$

(4 marks)

Question 2

State the order of the following matrices:

(a) $\begin{pmatrix} 2 & 5 \\ 2 & 6 \\ 3 & 9 \\ 1 & 2 \\ 3 & 2 \end{pmatrix}$ *number of rows × number of columns*

$\boxed{5 \times 2}$

(b) $\begin{pmatrix} 4 & 5 \\ 4 & 5 \end{pmatrix}$ *number of rows × number of columns*

$\boxed{2 \times 2}$

(c) $(2\ 3\ 4\ 6)$ *number of rows × number of columns*

$\boxed{1 \times 4}$

(d) $\begin{pmatrix} 3 \\ 2 \\ 3 \\ 6 \\ 7 \end{pmatrix}$ *number of rows × number of columns*

$\boxed{5 \times 1}$

(4 marks)

Question 3

Consider the following matrices:

$$A = \begin{pmatrix} 4 & 5 \\ 4 & 5 \end{pmatrix} \qquad B = \begin{pmatrix} 1 & 3 \\ 4 & 9 \end{pmatrix} \qquad C = \begin{pmatrix} 6 & 2 \\ 1 & 3 \end{pmatrix}$$

Determine the following:

(a) A + B =

$$\begin{pmatrix} 4 & 5 \\ 4 & 5 \end{pmatrix} + \begin{pmatrix} 1 & 3 \\ 4 & 9 \end{pmatrix} = \begin{pmatrix} 4+1 & 5+3 \\ 4+4 & 5+9 \end{pmatrix} = \boxed{\begin{pmatrix} 5 & 8 \\ 8 & 14 \end{pmatrix}}$$

(b) A + 2B =

$$A = \begin{pmatrix} 4 & 5 \\ 4 & 5 \end{pmatrix}$$

$$B = \begin{pmatrix} 1 & 3 \\ 4 & 9 \end{pmatrix}$$

$$2B = \begin{pmatrix} 2\times 1 & 2\times 3 \\ 2\times 4 & 2\times 9 \end{pmatrix} = \begin{pmatrix} 2 & 6 \\ 8 & 18 \end{pmatrix}$$

$$A + 2B = \begin{pmatrix} 4+2 & 5+6 \\ 4+8 & 5+18 \end{pmatrix} = \boxed{\begin{pmatrix} 6 & 11 \\ 12 & 23 \end{pmatrix}}$$

(c) 3B + A =

$$B = \begin{pmatrix} 1 & 3 \\ 4 & 9 \end{pmatrix}$$

$$3B = \begin{pmatrix} 3 \times 1 & 3 \times 3 \\ 3 \times 4 & 3 \times 9 \end{pmatrix} = \begin{pmatrix} 3 & 9 \\ 12 & 27 \end{pmatrix}$$

$$A = \begin{pmatrix} 4 & 5 \\ 4 & 5 \end{pmatrix}$$

$$3B + A = \begin{pmatrix} 3+4 & 9+5 \\ 12+4 & 27+5 \end{pmatrix} = \boxed{\begin{pmatrix} 7 & 14 \\ 16 & 32 \end{pmatrix}}$$

(d) 2A + 2B =

$$A = \begin{pmatrix} 4 & 5 \\ 4 & 5 \end{pmatrix} \quad 2A = \begin{pmatrix} 2 \times 4 & 2 \times 5 \\ 2 \times 4 & 2 \times 5 \end{pmatrix} = \begin{pmatrix} 8 & 10 \\ 8 & 10 \end{pmatrix}$$

$$B = \begin{pmatrix} 1 & 3 \\ 4 & 9 \end{pmatrix} \quad 2B = \begin{pmatrix} 2 \times 1 & 2 \times 3 \\ 2 \times 4 & 2 \times 9 \end{pmatrix} = \begin{pmatrix} 2 & 6 \\ 8 & 18 \end{pmatrix}$$

$$2A + 2B = \begin{pmatrix} 8+2 & 10+6 \\ 8+8 & 10+18 \end{pmatrix} = \boxed{\begin{pmatrix} 10 & 16 \\ 16 & 28 \end{pmatrix}}$$

(e) 2B + A - C =

$$2B = \begin{pmatrix} 2 & 6 \\ 8 & 18 \end{pmatrix} \quad A = \begin{pmatrix} 4 & 5 \\ 4 & 5 \end{pmatrix} \quad C = \begin{pmatrix} 6 & 2 \\ 1 & 3 \end{pmatrix}$$

$$2B + A - C = \begin{pmatrix} 2 & 6 \\ 8 & 18 \end{pmatrix} + \begin{pmatrix} 4 & 5 \\ 4 & 5 \end{pmatrix} - \begin{pmatrix} 6 & 2 \\ 1 & 3 \end{pmatrix}$$

$$2B + A - C = \begin{pmatrix} 2+4-6 & 6+5-2 \\ 8+4-1 & 18+5-3 \end{pmatrix}$$

$$2B + A - C = \boxed{\begin{pmatrix} 0 & 9 \\ 11 & 20 \end{pmatrix}}$$

(10 marks)

Question 4

Determine the values of P, Q, R and S.

(a) $\begin{pmatrix} 5 & P \\ 20 & 5 \end{pmatrix} + \begin{pmatrix} 3 & -5 \\ R & 6 \end{pmatrix} = \begin{pmatrix} Q & 10 \\ 18 & S \end{pmatrix}$

P + -5 = 10   20 + R = 18

P = 10 + 5 = 15   R = 18 - 20

| **P = 15** | | **R = -2** |

5 + 3 = Q   5 + 6 = S

| **Q = 8** | | **S = 11** |

Determine the values of A, B, C and D.

(b) $\begin{pmatrix} 3 & A \\ 4 & 9 \end{pmatrix} + \begin{pmatrix} 2 & 1 \\ 3 & 5 \end{pmatrix} = \begin{pmatrix} B & 3 \\ C & D \end{pmatrix}$

3 + 2 = B   4 + 3 = C

| **B = 5** | | **C = 7** |

A + 1 = 3   9 + 5 = D

A = 3 - 1 | **D = 14** |

| **A = 2** |

(4 marks)

Question 5

Consider the following matrices:

$$P = \begin{pmatrix} 1 & 2 \\ 2 & 6 \end{pmatrix} \quad Q = \begin{pmatrix} 9 & 10 \\ 8 & 7 \end{pmatrix} \quad R = \begin{pmatrix} 11 & 12 \\ 1 & 2 \end{pmatrix}$$

(a) $Q + R = \begin{pmatrix} 9 & 10 \\ 8 & 7 \end{pmatrix} + \begin{pmatrix} 11 & 12 \\ 1 & 2 \end{pmatrix} = \begin{pmatrix} 9+11 & 10+12 \\ 8+1 & 7+2 \end{pmatrix}$

$$\boxed{Q + R = \begin{pmatrix} 20 & 22 \\ 9 & 9 \end{pmatrix}}$$

(b) $P + 2Q + R =$

$P = \begin{pmatrix} 1 & 2 \\ 2 & 6 \end{pmatrix} \quad 2Q = \begin{pmatrix} 18 & 20 \\ 16 & 14 \end{pmatrix} \quad R = \begin{pmatrix} 11 & 12 \\ 1 & 2 \end{pmatrix}$

$P + 2Q + R = \begin{pmatrix} 1+18+11 & 2+20+12 \\ 2+16+1 & 6+14+2 \end{pmatrix} = \begin{pmatrix} 30 & 34 \\ 19 & 22 \end{pmatrix}$

$$\boxed{P + 2Q + R = \begin{pmatrix} 30 & 34 \\ 19 & 22 \end{pmatrix}}$$

(c) R - P =

$$R = \begin{pmatrix} 11 & 12 \\ 1 & 2 \end{pmatrix} \quad P = \begin{pmatrix} 1 & 2 \\ 2 & 6 \end{pmatrix}$$

$$R - P = \begin{pmatrix} 11-1 & 12-2 \\ 1-2 & 2-6 \end{pmatrix}$$

$$\boxed{R - P = \begin{pmatrix} 10 & 10 \\ -1 & -4 \end{pmatrix}}$$

(d) 2Q - R =

$$2Q = \begin{pmatrix} 2 \times 9 & 2 \times 10 \\ 2 \times 8 & 2 \times 7 \end{pmatrix} = \begin{pmatrix} 18 & 20 \\ 16 & 14 \end{pmatrix}$$

$$2Q = \begin{pmatrix} 18 & 20 \\ 16 & 14 \end{pmatrix}$$

$$2Q = \begin{pmatrix} 18 & 20 \\ 16 & 14 \end{pmatrix} \quad R = \begin{pmatrix} 11 & 12 \\ 1 & 2 \end{pmatrix}$$

$$2Q - R = \begin{pmatrix} 18-11 & 20-12 \\ 16-1 & 14-2 \end{pmatrix}$$

$$\boxed{2Q - R = \begin{pmatrix} 7 & 8 \\ 15 & 12 \end{pmatrix}}$$

(e) P + Q + R =

$$P = \begin{pmatrix} 1 & 2 \\ 2 & 6 \end{pmatrix} \quad Q = \begin{pmatrix} 9 & 10 \\ 8 & 7 \end{pmatrix} \quad R = \begin{pmatrix} 11 & 12 \\ 1 & 2 \end{pmatrix}$$

$$P + Q + R = \begin{pmatrix} 1+9+11 & 2+10+12 \\ 2+8+1 & 6+7+2 \end{pmatrix}$$

$$\boxed{P + Q + R = \begin{pmatrix} 21 & 24 \\ 11 & 15 \end{pmatrix}}$$

(10 marks)

Question 6

Consider the following matrices:

$$A = \begin{pmatrix} 1 & 5 \\ 2 & 6 \end{pmatrix} \quad B = \begin{pmatrix} 9 & 1 \\ 8 & 1 \end{pmatrix} \quad C = \begin{pmatrix} 11 & 12 \\ 1 & 1 \end{pmatrix}$$

(a) Find |A|, |B|, |C|

$A = \begin{pmatrix} 1 & 5 \\ 2 & 6 \end{pmatrix}$ let $A = \begin{pmatrix} a & b \\ c & d \end{pmatrix}$ |A| = (a)(d) – (b)(c)

|A| = (1)(6) – (5)(2) |A| = 6 - 10

|A| = -4

$B = \begin{pmatrix} 9 & 1 \\ 8 & 1 \end{pmatrix}$ let $B = \begin{pmatrix} a & b \\ c & d \end{pmatrix}$ |B| = (a)(d) – (b)(c)

|B| = (9)(1) – (1)(8)

|B| = 1

$C = \begin{pmatrix} 11 & 12 \\ 1 & 1 \end{pmatrix}$ let $C = \begin{pmatrix} a & b \\ c & d \end{pmatrix}$

|C| = (11)(1) – (12)(1)

|C| = 11-12

|C| = -1

(6 marks)

(b) Find A^{-1}

$A^{-1} = \frac{1}{|A|} \begin{pmatrix} d & -b \\ -c & a \end{pmatrix} = \frac{1}{(a)(d)-(b)(c)} \begin{pmatrix} d & -b \\ -c & a \end{pmatrix}$

$A^{-1} = -\frac{1}{4} \begin{pmatrix} 6 & -5 \\ -2 & 1 \end{pmatrix} \quad A^{-1} = \begin{pmatrix} -1.5 & 1.25 \\ 0.5 & -0.25 \end{pmatrix}$

(3 marks)

(c) Find B^{-1}

$B = \begin{pmatrix} 9 & 1 \\ 8 & 1 \end{pmatrix}$ let $B = \begin{pmatrix} a & b \\ c & d \end{pmatrix}$

$B^{-1} = \frac{1}{|B|} \begin{pmatrix} d & -b \\ -c & a \end{pmatrix} = \frac{1}{(a)(d)-(b)(c)} \begin{pmatrix} d & -b \\ -c & a \end{pmatrix}$

$B^{-1} = \frac{1}{1} \begin{pmatrix} 1 & -1 \\ -8 & 9 \end{pmatrix}$

$\boxed{B^{-1} = \begin{pmatrix} 1 & -1 \\ -8 & 9 \end{pmatrix}}$

(3 marks)

(d) Find C^{-1}

$C = \begin{pmatrix} 11 & 12 \\ 1 & 1 \end{pmatrix}$ let $C = \begin{pmatrix} a & b \\ c & d \end{pmatrix}$

$C^{-1} = \frac{1}{|C|} \begin{pmatrix} d & -b \\ -c & a \end{pmatrix} = \frac{1}{-1} \begin{pmatrix} 1 & -12 \\ -1 & 11 \end{pmatrix}$

$\boxed{C^{-1} = \begin{pmatrix} -1 & 12 \\ 1 & -11 \end{pmatrix}}$

(3 marks)

Question 7

Multiply the following matrices. If the matrices cannot be multiplied state a reason why.

(a) $\begin{pmatrix} 1 & 8 \\ 2 & 6 \end{pmatrix} \times \begin{pmatrix} 3 & 3 & 3 \\ 4 & 4 & 4 \end{pmatrix} = \begin{pmatrix} 1\times 3 + 8\times 4 & 1\times 3 + 8\times 4 & 1\times 3 + 8\times 4 \\ 2\times 3 + 6\times 4 & 2\times 3 + 6\times 4 & 2\times 3 + 6\times 4 \end{pmatrix}$

$\quad\quad 2\times 2 \quad\quad 2\times 3$

$\begin{pmatrix} 1\times 3 + 8\times 4 & 1\times 3 + 8\times 4 & 1\times 3 + 8\times 4 \\ 2\times 3 + 6\times 4 & 2\times 3 + 6\times 4 & 2\times 3 + 6\times 4 \end{pmatrix} = \boxed{\begin{pmatrix} 35 & 35 & 35 \\ 30 & 30 & 30 \end{pmatrix}}$

(2 marks)

(b) $\begin{pmatrix} 9 & 10 \\ 8 & 9 \end{pmatrix} \times \begin{pmatrix} 11 & 12 \\ 1 & 1 \end{pmatrix} = \begin{pmatrix} 9\times 11 + 10\times 1 & 9\times 12 + 10\times 1 \\ 8\times 11 + 9\times 1 & 8\times 12 + 9\times 1 \end{pmatrix}$

$\quad 2\times 2 \quad\quad 2\times 2 \quad = \begin{pmatrix} 99+10 & 108+10 \\ 88+9 & 96+9 \end{pmatrix} = \boxed{\begin{pmatrix} 109 & 118 \\ 97 & 105 \end{pmatrix}}$

(2 marks)

c) $\begin{pmatrix} 3 & 3 & 3 \\ 4 & 4 & 4 \end{pmatrix} \times \begin{pmatrix} 1 & 8 \\ 2 & 6 \end{pmatrix} =$

$\quad 2\times 3 \quad\quad 2\times 2$

These matrices cannot be multiplied since the number of columns in the first matrix is not equal to number of rows in the second matrix.

(2 marks)

Question 8

$$P = \begin{pmatrix} 1 & 3 \\ 1 & 6 \end{pmatrix} \quad Q = \begin{pmatrix} 9 & 16 \\ 1 & 2 \end{pmatrix} \quad R = \begin{pmatrix} 11 & 11 \\ 0 & 1 \end{pmatrix}$$

(a) Find PQ

$$PQ = \begin{pmatrix} 1 & 3 \\ 1 & 6 \end{pmatrix} \times \begin{pmatrix} 9 & 16 \\ 1 & 2 \end{pmatrix} = \begin{pmatrix} 1 \times 9 + 3 \times 1 & 1 \times 16 + 3 \times 2 \\ 1 \times 9 + 6 \times 1 & 1 \times 16 + 6 \times 2 \end{pmatrix}$$

$$PQ = \begin{pmatrix} 9+3 & 16+6 \\ 9+6 & 16+12 \end{pmatrix} \quad \boxed{PQ = \begin{pmatrix} 12 & 22 \\ 15 & 28 \end{pmatrix}}$$

(2 marks)

(b) Find QR

$$QR = \begin{pmatrix} 9 & 16 \\ 1 & 2 \end{pmatrix} \times \begin{pmatrix} 11 & 11 \\ 0 & 1 \end{pmatrix} = \begin{pmatrix} 9 \times 11 + 16 \times 0 & 9 \times 11 + 16 \times 1 \\ 1 \times 11 + 2 \times 0 & 1 \times 11 + 2 \times 1 \end{pmatrix}$$

$$QR = \begin{pmatrix} 99+0 & 99+16 \\ 11+0 & 11+2 \end{pmatrix} \quad \boxed{QR = \begin{pmatrix} 99 & 115 \\ 11 & 13 \end{pmatrix}}$$

(2 marks)

(c) Find RQ

$RQ = \begin{pmatrix} 11 & 11 \\ 0 & 1 \end{pmatrix} \times \begin{pmatrix} 9 & 16 \\ 1 & 2 \end{pmatrix} = \begin{pmatrix} 11 \times 9 + 11 \times 1 & 11 \times 16 + 11 \times 2 \\ 0 \times 9 + 1 \times 1 & 0 \times 16 + 1 \times 2 \end{pmatrix}$

$RQ = \begin{pmatrix} 99 + 11 & 176 + 22 \\ 0 + 1 & 0 + 2 \end{pmatrix}$

$\boxed{RQ = \begin{pmatrix} 110 & 198 \\ 1 & 2 \end{pmatrix}}$

(2 marks)

(d) Find PQR

$PQR = PQ \times R$

Recall from (a) $PQ = \begin{pmatrix} 12 & 22 \\ 15 & 28 \end{pmatrix}$

$PQR = \begin{pmatrix} 12 & 22 \\ 15 & 28 \end{pmatrix} \times \begin{pmatrix} 11 & 11 \\ 0 & 1 \end{pmatrix}$

$PQR = \begin{pmatrix} 12 \times 11 + 22 \times 0 & 12 \times 11 + 22 \times 1 \\ 15 \times 11 + 28 \times 0 & 15 \times 11 + 28 \times 1 \end{pmatrix}$

$PQR = \begin{pmatrix} 132 + 0 & 132 + 22 \\ 165 + 0 & 165 + 28 \end{pmatrix}$

$\boxed{PQR = \begin{pmatrix} 132 & 154 \\ 165 & 193 \end{pmatrix}}$

(2 marks)

Question 9

(a) Express the equations

2x - 6y = 0

3x + 2y = 44

in the form AX = B, where A, X and B are matrices.

$$\begin{pmatrix} 2 & -6 \\ 3 & 2 \end{pmatrix} \begin{pmatrix} x \\ y \end{pmatrix} = \begin{pmatrix} 0 \\ 44 \end{pmatrix}$$

$$A = \begin{pmatrix} 2 & -6 \\ 3 & 2 \end{pmatrix} \quad X = \begin{pmatrix} x \\ y \end{pmatrix} \quad B = \begin{pmatrix} 0 \\ 44 \end{pmatrix}$$

(2 marks)

(b) Express the equations

4x - 3y = 5

5x - 2y = 8

in the form AX = B, where A, X and B are matrices.

$$\begin{pmatrix} 4 & -3 \\ 5 & -2 \end{pmatrix} \begin{pmatrix} x \\ y \end{pmatrix} = \begin{pmatrix} 5 \\ 8 \end{pmatrix}$$

$$A = \begin{pmatrix} 4 & -3 \\ 5 & -2 \end{pmatrix} \quad X = \begin{pmatrix} x \\ y \end{pmatrix} \quad B = \begin{pmatrix} 5 \\ 8 \end{pmatrix}$$

(2 marks)

Question 10

Solve the simultaneous equations using matrices:

x + y = 14

2x + 3y = 33

$$\begin{pmatrix} 1 & 1 \\ 2 & 3 \end{pmatrix} \begin{pmatrix} x \\ y \end{pmatrix} = \begin{pmatrix} 14 \\ 33 \end{pmatrix}$$

   A   **X**   **B**

X = A⁻¹ B

$$\begin{pmatrix} x \\ y \end{pmatrix} = \frac{1}{|A|} \begin{pmatrix} 3 & -1 \\ -2 & 1 \end{pmatrix} \begin{pmatrix} 14 \\ 33 \end{pmatrix}$$

let $A = \begin{pmatrix} a & b \\ c & d \end{pmatrix}$

|A| = (a)(d) − (b)(c) |A| = (1)(3)−(1)(2) |A| = 3−2 |A| =1

$$\begin{pmatrix} x \\ y \end{pmatrix} = \frac{1}{1} \begin{pmatrix} 3 & -1 \\ -2 & 1 \end{pmatrix} \begin{pmatrix} 14 \\ 33 \end{pmatrix} \qquad \begin{pmatrix} x \\ y \end{pmatrix} = \begin{pmatrix} 3 & -1 \\ 2 & 1 \end{pmatrix} \begin{pmatrix} 14 \\ 33 \end{pmatrix}$$

                                               2×2   2×1

$$\begin{pmatrix} x \\ y \end{pmatrix} = \begin{pmatrix} 3 & -1 \\ -2 & 1 \end{pmatrix} \begin{pmatrix} 14 \\ 33 \end{pmatrix}$$

$$\begin{pmatrix} x \\ y \end{pmatrix} = \begin{pmatrix} 3 \times 14 & + & -1 \times 33 \\ -2 \times 14 & + & 1 \times 33 \end{pmatrix}$$

$$\begin{pmatrix} x \\ y \end{pmatrix} = \begin{pmatrix} 42 - 33 \\ -28 + 33 \end{pmatrix}$$

$$\begin{pmatrix} x \\ y \end{pmatrix} = \begin{pmatrix} 9 \\ 5 \end{pmatrix}$$

x = 9
y = 5

(6 marks)

Question 11

Solve the simultaneous equations using matrices.

a + b = 10

3a + 2b = 28

$$\begin{pmatrix} 1 & 1 \\ 3 & 2 \end{pmatrix} \begin{pmatrix} a \\ b \end{pmatrix} = \begin{pmatrix} 10 \\ 28 \end{pmatrix}$$

  A   X   B

$X = A^{-1} B$

$\begin{pmatrix} a \\ b \end{pmatrix} = \frac{1}{|A|} \begin{pmatrix} 2 & -1 \\ -3 & 1 \end{pmatrix} \begin{pmatrix} 10 \\ 28 \end{pmatrix}$ let $A = \begin{pmatrix} a & b \\ c & d \end{pmatrix}$

$|A| = (a)(d) - (b)(c)$ $|A| = (1)(2) - (1)(3)$ $|A| = 2-3$ $|A| = -1$

$\begin{pmatrix} a \\ b \end{pmatrix} = \frac{1}{-1} \begin{pmatrix} 2 & -1 \\ -3 & 1 \end{pmatrix} \begin{pmatrix} 10 \\ 28 \end{pmatrix}$ $\begin{pmatrix} a \\ b \end{pmatrix} = \begin{pmatrix} -2 & 1 \\ 3 & -1 \end{pmatrix} \begin{pmatrix} 10 \\ 28 \end{pmatrix}$

                                                                             2×2   2×1

$\begin{pmatrix} a \\ b \end{pmatrix} = \begin{pmatrix} -2 & 1 \\ 3 & -1 \end{pmatrix} \begin{pmatrix} 10 \\ 28 \end{pmatrix}$

$\begin{pmatrix} a \\ b \end{pmatrix} = \begin{pmatrix} -2 \times 10 & + & 1 \times 28 \\ 3 \times 10 & + & -1 \times 28 \end{pmatrix}$

$\begin{pmatrix} a \\ b \end{pmatrix} = \begin{pmatrix} -20 + 28 \\ 30 - 28 \end{pmatrix}$

$\begin{pmatrix} a \\ b \end{pmatrix} = \begin{pmatrix} 8 \\ 2 \end{pmatrix}$

---

a = 8

b = 2

(6 marks)

Question 12

Solve the simultaneous equations using matrices.

p − q = 10

2p + q = 26

$$\begin{pmatrix} 1 & -1 \\ 2 & 1 \end{pmatrix} \begin{pmatrix} p \\ q \end{pmatrix} = \begin{pmatrix} 10 \\ 26 \end{pmatrix}$$

   A   X   B

$X = A^{-1} B$

$$\begin{pmatrix} p \\ q \end{pmatrix} = \frac{1}{|A|} \begin{pmatrix} 1 & 1 \\ -2 & 1 \end{pmatrix} \begin{pmatrix} 10 \\ 26 \end{pmatrix} \quad \text{let } A = \begin{pmatrix} a & b \\ c & d \end{pmatrix}$$

$|A| = (a)(d) - (b)(c) \quad |A| = (1)(1) - (-1)(2) \quad |A| = 1 + 2 \quad |A| = 3$

$$\begin{pmatrix} p \\ q \end{pmatrix} = \frac{1}{3} \begin{pmatrix} 1 & 1 \\ -2 & 1 \end{pmatrix} \begin{pmatrix} 10 \\ 26 \end{pmatrix} \quad \begin{pmatrix} a \\ b \end{pmatrix} = \begin{pmatrix} \frac{1}{3} & \frac{1}{3} \\ \frac{-2}{3} & \frac{1}{3} \end{pmatrix} \begin{pmatrix} 10 \\ 26 \end{pmatrix}$$

                                                      2×②   ②×1

$$\begin{pmatrix} p \\ q \end{pmatrix} = \begin{pmatrix} \frac{1}{3} \times 10 + \frac{1}{3} \times 26 \\ \frac{-2}{3} \times 10 + \frac{1}{3} \times 26 \end{pmatrix}$$

$$\begin{pmatrix} p \\ q \end{pmatrix} = \begin{pmatrix} \frac{10}{3} + \frac{26}{3} \\ \frac{-20}{3} + \frac{26}{3} \end{pmatrix}$$

$$\begin{pmatrix} p \\ q \end{pmatrix} = \begin{pmatrix} \frac{36}{3} \\ \frac{6}{3} \end{pmatrix}$$

$$\begin{pmatrix} p \\ q \end{pmatrix} = \begin{pmatrix} 12 \\ 2 \end{pmatrix}$$

p = 12
q = 2

(6 marks)

Question 13

(a) The Matrix A is defined as:

$$A = \begin{pmatrix} 5 & 2 \\ b & 4 \end{pmatrix}$$

Determine the value of b for which the matrix A does not have an inverse.

Let $A = \begin{pmatrix} w & x \\ y & z \end{pmatrix}$ $\quad |A| = (w)(z) - (x)(y)$

The matrix A does not have an inverse when $|A| = 0$

$0 = (5)(4) - (2)(b)$

$20 - 2b = 0$

$-2b = -20$

$b = \frac{20}{2}$ $\quad \boxed{b = 10}$ *(The matrix A does not have an inverse when b=10)*

(2 marks)

(b) The Matrix C is defined as:

$$C = \begin{pmatrix} 6 & d \\ 2 & 4 \end{pmatrix}$$

Determine the value of d for which the matrix C does not have an inverse.

Let $C = \begin{pmatrix} w & x \\ y & z \end{pmatrix}$ $\quad |A| = (w)(z) - (x)(y)$

The matrix C does not have an inverse when $|C| = 0$

$0 = (6)(4) - (d)(2)$

$24 - 2d = 0$

$-2d = -24$

$d = \frac{-24}{-2}$ $\quad \boxed{d = 12}$ *(The matrix C does not have an inverse when d =12)*

(2 marks)

Question 14

(a) The Matrix Y is defined as:

$$Y = \begin{pmatrix} -5 & 2 \\ p & 4 \end{pmatrix}$$

Determine the value of p for which the matrix Y does not have an inverse.

Let $Y = \begin{pmatrix} a & b \\ c & d \end{pmatrix}$ $|Y| = (a)(d) - (b)(c)$

The matrix A does not have an inverse when $|Y| = 0$

$0 = (-5)(4) - (2)(p)$

$-20 - 2p = 0$

$-2p = -20$

$p = \frac{20}{2}$ $\boxed{p = 10}$ (The matrix A does not have an inverse when $p = 10$)

(2 marks)

(b) The Matrix Z is defined as:

$$Z = \begin{pmatrix} 10 & h \\ 9 & 18 \end{pmatrix}$$

Determine the value of h for which the matrix Z does not have an inverse.

Let $Z = \begin{pmatrix} a & b \\ c & d \end{pmatrix}$ $|Z| = (a)(d) - (b)(c)$

The matrix A does not have an inverse when $|Z| = 0$

$0 = (10)(18) - (h)(9)$

$180 - 9h = 0$

$-9h = -180$

$h = \frac{-180}{-9}$ $\boxed{h = 20}$ (The matrix A does not have an inverse when $h = 20$)

(2 marks)

Question 15

(a) State the 2 × 2 transformation matrix which represents a reflection in the line y = x.

$\begin{pmatrix} 0 & 1 \\ 1 & 0 \end{pmatrix}$

(b) State the 2 × 2 transformation matrix which represents a reflection in the line y = -x.

$\begin{pmatrix} 0 & -1 \\ -1 & 0 \end{pmatrix}$

(c) State the 2 × 2 transformation matrix which represents a reflection in the x axis.

$\begin{pmatrix} 1 & 0 \\ 0 & -1 \end{pmatrix}$

(d) State the 2 × 2 transformation matrix which represents a reflection in the y axis.

$\begin{pmatrix} -1 & 0 \\ 0 & 1 \end{pmatrix}$

(e) State the 2 × 2 transformation matrix which represents a 90° clockwise rotation about the origin.

$\begin{pmatrix} 0 & 1 \\ -1 & 0 \end{pmatrix}$

(10 marks)

END OF WORKSHEET

868

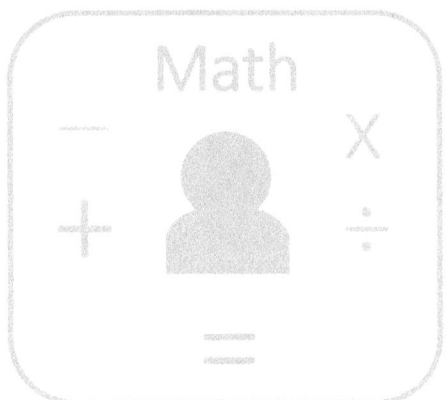

Preparation for

High School Mathematics

Measurement

(Speed, Distance, Time)

Solutions

Instructions and Tips:

- ✓ **You have 60 minutes to complete this worksheet**
- ✓ **This worksheet consists of 5 questions**
- ✓ **Write answers in the spaces provided**
- ✓ **All working must be clearly shown**
- ✓ **Give answers to 2 decimal places**

Student Name: _____

Student ID: _____

Date: __/__/____

Total Score:

Highest Score:

Tutor's Comments:

Access more free worksheets at www.868tutors.com

Question 1

A sprinter from St. Kitts and Nevis enters a regional 100 m race. The sprinter covers a distance of 100 m in 9.87 seconds.

(a) Calculate his average speed for the race.

Average speed = $\frac{Total\ Distance}{Total\ Time}$ = $\frac{100\ m}{9.87\ seconds}$

Average speed = $\boxed{10.13\ m/s\ \text{(to 2 decimal places)}}$

(2 marks)

(b) What should his average speed be to complete the 100 m in 9.59 seconds?

Average speed $_{required}$ = $\frac{100\ m}{9.59\ s}$ = $\boxed{10.43\ m/s\ \text{(to 2 decimal places)}}$

(2 marks)

(c) The sprinter's average speed for a 200 m race is 10.21 m/s. Calculate the time he takes to cover 200 m.

Average speed = $\frac{Total\ Distance}{Total\ Time}$ Time × Average Speed = Total Distance

Time = $\frac{Total\ Distance}{Average\ Speed}$ Time = $\frac{200\ m}{10.21\ m/s}$ Time = $\boxed{19.59\ \text{seconds (to 2 decimal places)}}$

(2 marks)

Question 2

A pickup truck travels at 65 kilometres per hour on a highway in Trinidad.

(a) Calculate the time taken, in minutes, to cover 30 km at this constant speed.

Average speed = $\frac{Total\ Distance}{Total\ Time}$ Time = $\frac{Total\ Distance}{Average\ Speed} = \frac{30\ km}{65\ km/h}$

Time = 0.461538462 hour

1 hour = 60 minutes   0.461538462 hour = 60 minutes × 0.461538462 = **27.69 minutes (to 2 dp)**

(1 mark)

(b) Calculate the time taken, in minutes, to cover 30 km at a constant speed of 80 km/h.

Average speed = $\frac{Total\ Distance}{Total\ Time}$ Time = $\frac{Total\ Distance}{Average\ Speed} = \frac{30\ km}{80\ km/h}$

Time = 0.375 hour

1 hour = 60 minutes   0.375 hour = 60 minutes × 0.375 = **22.5 minutes (to 2 dp)**

(2 marks)

(c) How much time is saved by travelling at 80 km/h instead of 65 km/h over a distance of 30 km?

Time saved = Time taken at 65km/h − Time taken at 80 km/h

Time saved = 27.69 minutes − 22.5 minutes

Time saved = **5.19 minutes (to 2 dp)**

(2 marks)

Question 3

The chart below shows the 2 kilometre sprint times of some speed boats in a race off the coast of Trinidad.

Name of Speedboat	Time
Icacos Fire	51 seconds
Erin Dragon	50 seconds
St. Patrick Speedster	49 seconds
Mr. La Brea	45 seconds

(a) Calculate the average speed (in km/h) of Mr. La Brea during the sprint.

Average speed = $\frac{Total\ Distance}{Total\ Time}$ = $\frac{2\ km}{Total\ Time}$ Total Time (Mr. La Brea) = 45 seconds

3600 seconds = 1 hour   1 second = $\frac{1\ hour}{3600}$ 45 seconds = $\frac{1\ hour}{3600} \times 45$ 45 seconds = 0.0125 hour

Average speed = $\frac{Total\ Distance}{Total\ Time}$ = $\frac{2\ km}{0.0125\ hour}$ = $\boxed{160\ km/h}$ **(2 marks)**

(b) Calculate the average speed (in km/h) of Icacos Fire during the sprint.

Average speed = $\frac{Total\ Distance}{Total\ Time}$ = $\frac{2\ km}{Total\ Time}$ Total Time (Icacos Fire) = 51 seconds

3600 seconds = 1 hour   1 second = $\frac{1\ hour}{3600}$ 51 seconds = $\frac{1\ hour}{3600} \times 51$ 51 seconds = 0.014166666 hr

Average speed = $\frac{Total\ Distance}{Total\ Time}$ = $\frac{2\ km}{0.014166667\ hr}$ = $\boxed{141.18\ km/h\ (to\ 2\ dp)}$ **(2 marks)**

(c) The crew of Mr. La Brea want to achieve a time of 43 seconds in the next race. What should their average speed in km/h be?

Average speed = $\frac{Total\ Distance}{Total\ Time}$ = $\frac{2\ km}{Total\ Time}$ Required Time (Mr. La Brea) = 43 seconds

3600 seconds = 1 hour   1 second = $\frac{1\ hour}{3600}$ 43 seconds = $\frac{1\ hour}{3600} \times 43$ 43 seconds = 0.011944444 hour

Average speed = $\frac{Total\ Distance}{Total\ Time}$ = $\frac{2\ km}{0.011944444\ hour}$ = $\boxed{167.44\ km/h\ (to\ 2\ dp)}$

(3 marks)

Question 4

The table below indicates the times recorded by five horses at a horse racing event, on the beach, in Cedros, Trinidad. The race distance is 1.5 km.

Name of Horse	Time
Palo Seco Spirit	72 seconds
Apache	73 seconds
Survivor	74 seconds
Coromandel Commander	68 seconds

(a) Calculate the average speed (in m/s) of each horse.

Average speed = $\frac{Total\ Distance}{Total\ Time}$ 1 km = 1000 m   1.5 km = 1000 m × 1.5   1.5 km = 1500 m

Average Speed (Palo Seco Spirit) = $\frac{1500\ m}{72\ seconds}$ = **20.83 m/s (to 2 dp)**

Average Speed (Apache) = $\frac{1500\ m}{73\ seconds}$ = **20.55 m/s (to 2 dp)**

Average Speed (Survivor) = $\frac{1500\ m}{74\ seconds}$ = **20.27 m/s (to 2 dp)**

Average Speed (Coromandel Commander) = $\frac{1500\ m}{68\ seconds}$ = **22.06 m/s (to 2 dp)**

(2 marks)

(b) Complete the table below to show the position of each horse in the race.

Place	Name of Horse
1st	Coromandel Commander
2nd	Palo Seco Spirit
3rd	Apache
4th	Survivor

(2 marks)

Question 5

A ferry leaves Port of Spain with passengers headed for a family day at Columbus Bay. The distance to be covered is 80 km.

(a) Calculate the average speed (in km/h) the ferry has to travel to arrive in 1 hour and 15 minutes.

Average speed = $\frac{Total\ Distance}{Total\ Time}$ 1 hour and 15 minutes = 1.25 hours

Average speed = $\frac{80\ km}{1.25}$ = $\boxed{64\ \text{km/h}}$

(2 marks)

(b) The ferry leaves Port of Spain at 6am but arrives at its destination at 7:30 am. Calculate the average speed of the ferry in this case.

Average speed = $\frac{Total\ Distance}{Total\ Time}$ Time taken = 1 hour and 30 minutes = 1.5 hours

Average speed = $\frac{80\ km}{1.5\ hour}$ = $\boxed{53.33\ \text{km/h (to 2 dp)}}$

(2 marks)

END OF WORKSHEET

Access more free worksheets at www.868tutors.com

868

868TUTORS

TUTORS

Preparation for

High School Mathematics

Measurement II

Solutions

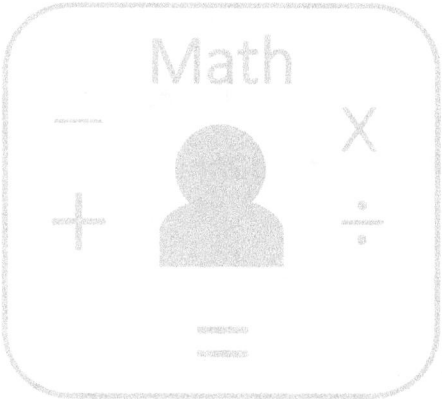

Instructions and Tips:

- ✓ You have 60 minutes to complete this worksheet
- ✓ This worksheet consists of 5 questions
- ✓ Write answers in the spaces provided
- ✓ All working must be clearly shown

Student Name: _____

Student ID: _____

Date: __ / __ / ____

Total Score:

Highest Score:

Tutor's Comments:

Access more free worksheets at www.868tutors.com

Question 1

Use π = 3.14

(a) Consider a rectangular room with a length of 20 m and a width of 10 m. Calculate the area of carpet that needs to be purchased to carpet the room.

Area = length × width

Area = 20m × 10m

$\boxed{\text{Area = 200 m}^2}$

(2 marks)

(b) Calculate the radius of a sphere that has a volume of 1000m³.

Volume of a sphere $= \frac{4}{3}\pi r^3$ $V = \frac{4}{3}\pi r^3$

Make r the subject of the formula $r = \left(\frac{0.75\,V}{\pi}\right)^{\frac{1}{3}}$

$r = \left(\frac{0.75 \times 1000}{\pi}\right)^{\frac{1}{3}}$ $r = \left(\frac{750}{\pi}\right)^{\frac{1}{3}}$ $r = \left(\frac{750}{3.14}\right)^{\frac{1}{3}}$ $\boxed{\text{radius = 6.20 m (to 2 decimal places)}}$

(2 marks)

(c) Calculate the surface area of a sphere that has a volume of 1000m³.

A (Surface area of a sphere) = $4\pi r^2$ recall from (b) that r = $\left(\frac{750}{\pi}\right)^{\frac{1}{3}}$ *using r = 6.203504909*

A = $4\pi r^2$ A = 4(3.14)(38.48347316) $\boxed{\text{A = 483.35 m}^2 \text{ (to 2 decimal places)}}$

(2 marks)

(d) Calculate the volume of a pyramid that has a base area of 20m² and a height of 5m.

Volume of a pyramid = Area of base × height

Volume of pyramid = 20m² × 5m

$\boxed{\text{Volume of pyramid = 100 m}^3}$

(2 marks)

Question 2

Consider the island below. The map is drawn on a grid of 1 cm squares. A, B, C, D and E are five high producing oil facilities.

The scale of the map is 1:2500

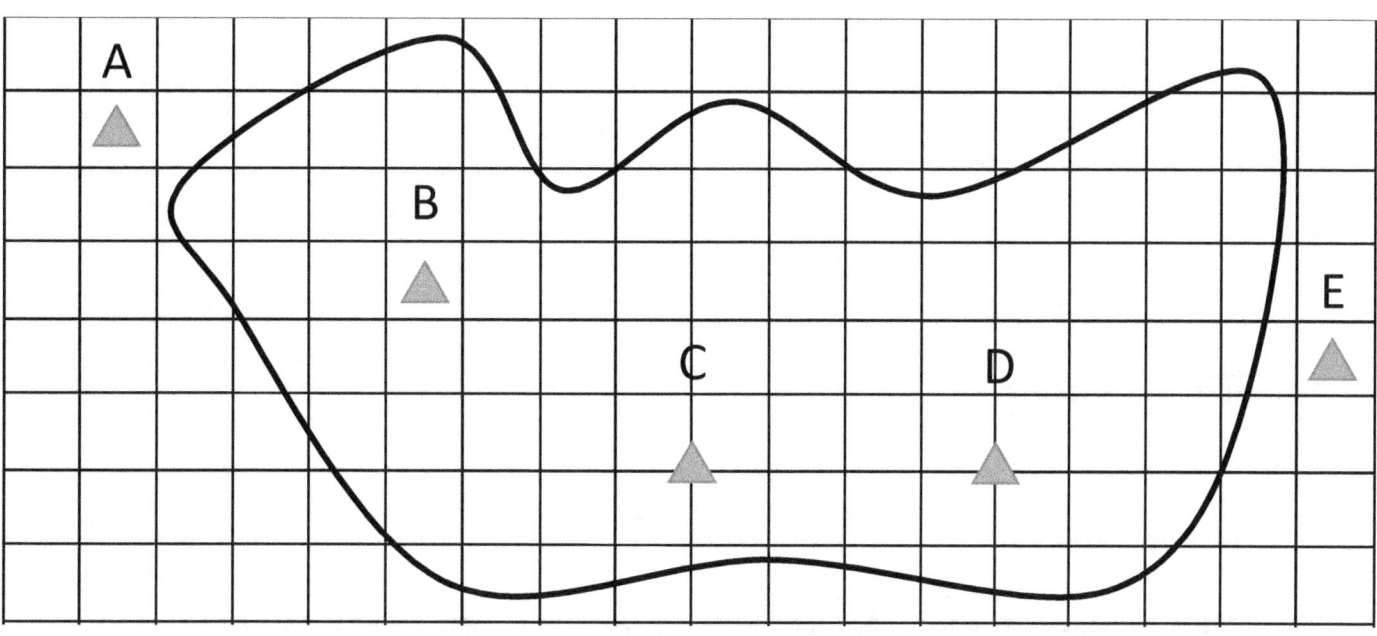

(a) Determine, in centimetres, the distance from C to D on the map.

C to D = 4 cm

(1 mark)

(b) Estimate, by counting, the area in square centimetres of the island.

Estimate = 76 cm²

(2 marks)

(c) Use the scale to Calculate the ACTUAL distance between C and D in kilometres on the map.

Scale 1: 2500

1 cm = 2,500 cm

4 cm = 10,000 cm

Actual distance = 10,000 cm

Converting cm to m

100 cm = 1 m

1 cm = $\frac{1}{100}$ m

10,000 cm = $\frac{1}{100}$ m × 10,000 = 100 m

Converting m to km

1000 m = 1 km

1 m = $\frac{1}{1000}$ km   100 m = $\frac{1}{1000}$ km × 100 = 0.1 km

Actual distance between C and D = 0.1 km

(2 marks)

(d) Calculate, the ACTUAL area in square metres of the island.

Estimate = 76 cm²

Converting estimate using scale

Scale 1:2500

1 cm = 2500 cm   1 cm² = (2500 cm)²

1 cm² = 6,250,000 cm²

76 cm² = 6,250,000 cm² × 76

76 cm² = 6,250,000 cm² × 76 = 475,000,000 cm²

Converting to m²

100 cm = 1 m   1 cm = $\frac{1}{100}$ m   1 cm² = $\left(\frac{1}{100} m\right)^2$ 475,000,000 cm² = $\left(\frac{1}{100} m\right)^2$ × 475,000,000

Actual area = 47,500 m²

(3 marks)

Question 3

Consider the island below. The map is drawn on a grid of 1 cm squares. X, Y and Z are three all-inclusive tourist resorts.

The scale of the map is 1:1500

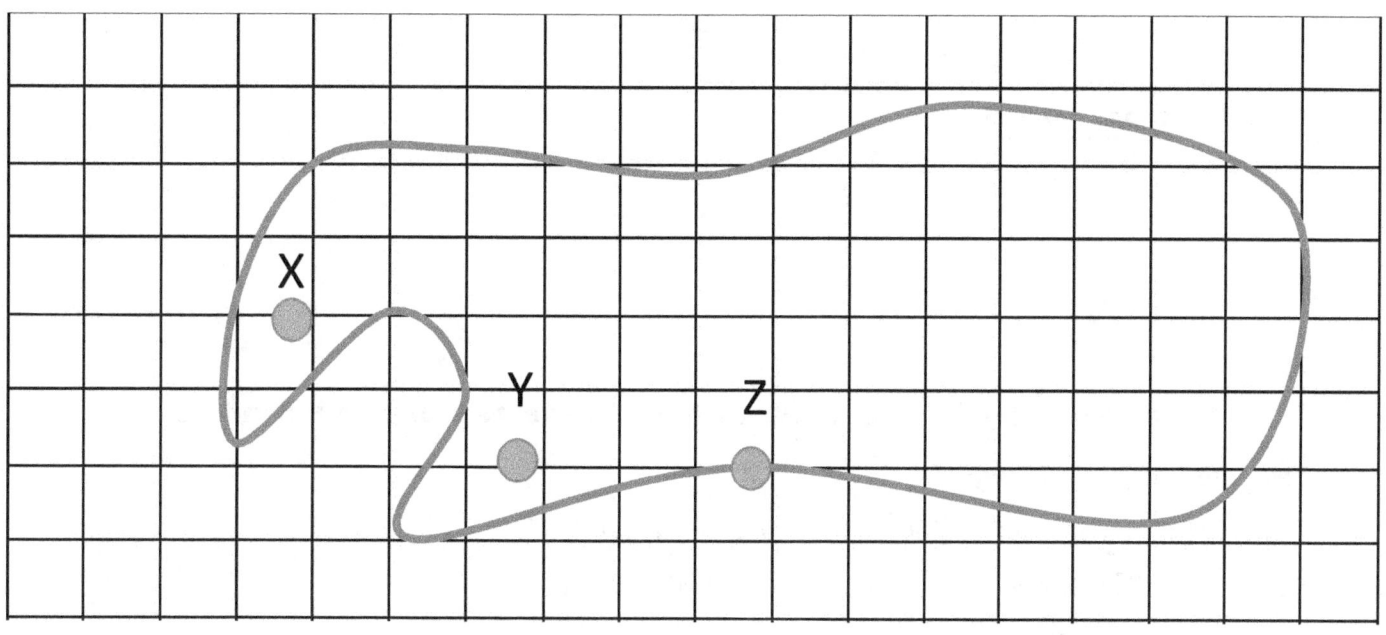

(a) Determine, in centimetres, the distance from Y to Z on the map.

Y to Z = 3 cm

(1 mark)

(b) Estimate, by counting, the area in square centimetres of the island.

Estimate = 54 cm^2

(2 marks)

(c) Use the scale to Calculate the ACTUAL distance in kilometres between Y and Z on the map.

Scale 1:1500   1 cm = 1500 cm   3 cm = 1500 cm × 3   3 cm = 4500 cm

Converting 4500 cm to m then to km

100 cm = 1 m

1 cm = $\frac{1}{100}$ m   4500 cm = $\frac{1}{100}$ m × 4500   4500 cm = 45 m

1000 m = 1 km

1 m = $\frac{1}{1000}$ km   45 m = $\frac{1}{1000}$ km × 45 = 0.045 km

Actual distance = 0.045 km

(3 marks)

(d) Calculate, the ACTUAL area in square metres of the island.

Estimate from (b) = 54 cm²

1 cm = 1500 cm   1 cm² = (1500 cm)²   1 cm² = 2,250,000 cm²

54 cm² = 2,250,000 cm² × 54   54 cm² = 121,500,000 cm²

Converting 121,500,000 cm² to m²

100 cm = 1m   1 cm = $\frac{1}{100}$ m   1 cm² = $\left(\frac{1}{100} m\right)^2$

121,500,000 cm² = $\left(\frac{1}{100} m\right)^2$ × 121,500,000 cm² = 12,150 m²

Actual area = 12,150 m²

(4 marks)

Question 4

Consider the cube below. The cube has a volume of 100 m³.

(Diagram not drawn to scale)

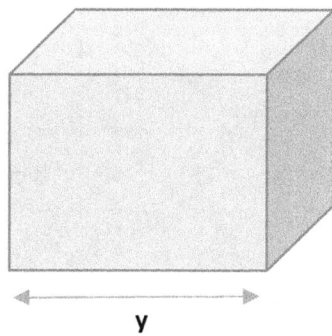

y

(a) Calculate the length of one side of the cube.

Volume of a cube = side × side × side

$V = y^3$ $100\ m = y^3$ $y = (100)^{\frac{1}{3}}$ length of one side = 4.64 m (to 2 decimal places)

(2 marks)

(b) Calculate the surface area of the cube.

Surface area of the cube = 6 × side² = 6y² = 6 × (4.641588834)²

Surface area of the cube = 129.27 m² (to 2 decimal places)

(2 marks)

Question 5

Consider the dimensions of the cuboid shown:

(Diagram not drawn to scale)

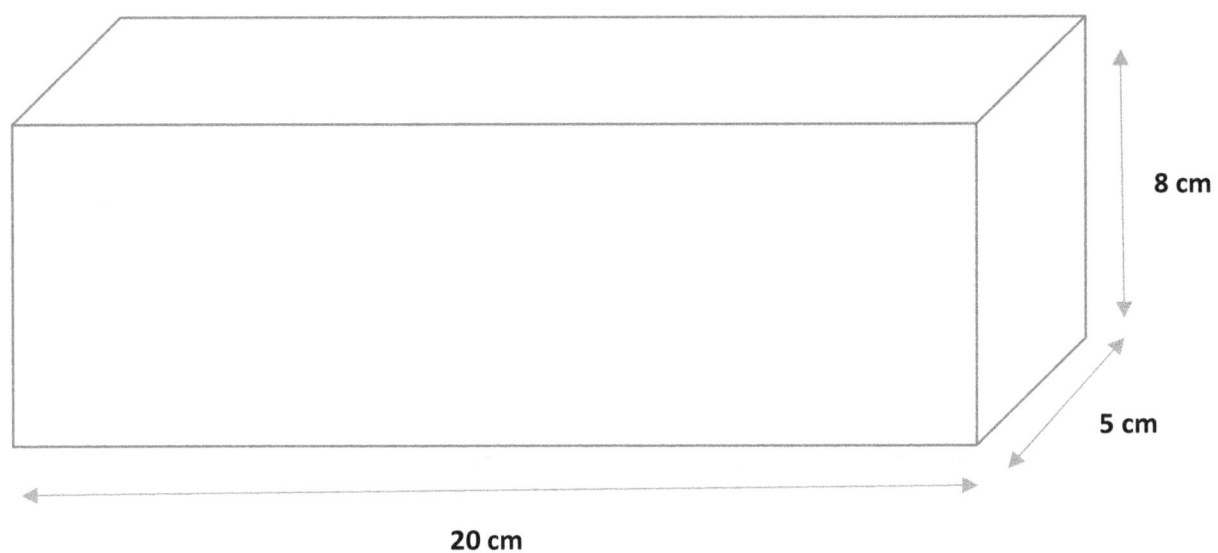

(a) Calculate the volume of the cuboid shown.

Volume of a cuboid = length × width × height

Volume of the cuboid = 20 cm × 5 cm × 8 cm

Volume of the cuboid = 800 cm³

(2 marks)

(b) Calculate the surface area of the cuboid shown.

Surface area of cuboid = (2 × length × height) + (2 × width × height) + (2 × length × width)

Surface area of cuboid = (2 × 20 cm × 8 cm) + (2 × 5 cm × 8 cm) + (2 × 20 cm × 5 cm)

Surface area of cuboid = (320 + 80 + 200) cm²

(2 marks)

Surface area of cuboid = 600 cm²

END OF WORKSHEET

868

868TUTORS

TUTORS

Preparation for

High School Mathematics

Measurement III

Solutions

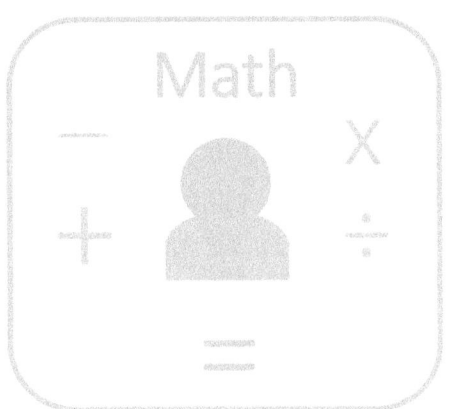

Instructions and Tips:

- ✓ You have 75 minutes to complete this worksheet
- ✓ This worksheet consists of 12 questions
- ✓ Write answers in the spaces provided
- ✓ All working must be clearly shown
- ✓ Diagrams are not drawn to scale

Student Name: _____

Student ID: _____

Date: __ / __ / ____

Total Score:

Highest Score:

Tutor's Comments:

Access more free worksheets at www.868tutors.com

Question 1

Consider the circle below with centre A and a radius of 6 cm:

Use π = 3.14

(Diagram not drawn to scale)

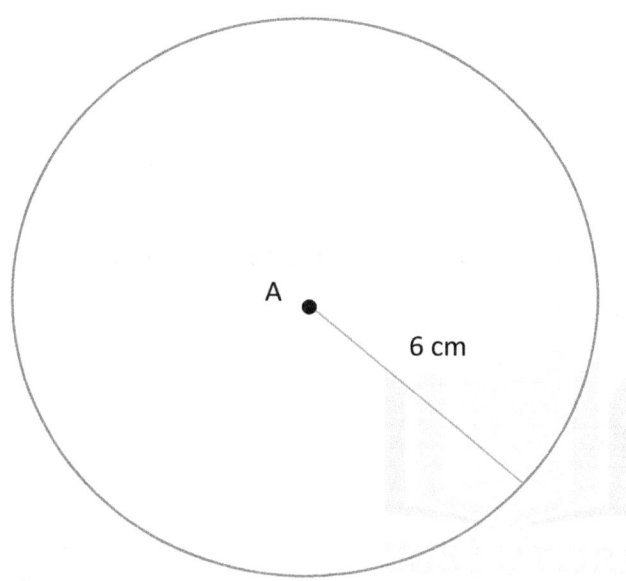

(a) Calculate the area of the circle.

$A = \pi r^2$
$A = 3.14 \times (6\ cm)^2$
$A = 3.14 \times 36\ cm^2$
Area of the circle = $\boxed{113.04\ \text{cm}^2}$

(2 marks)

(b) Calculate the circumference of the circle.

$C = 2\pi r$

$C = 2 \times 3.14 \times 6\ cm$

Circumference of the circle = $\boxed{37.68\ \text{cm}}$

(2 marks)

Question 2

Use π = 3.14

(a) Calculate the area of a circle of diameter 5 m.

$A = \pi r^2$ diameter = 5m   radius = 2.5 m

$A = 3.14 \times (2.5)^2$

$A = 3.14 \times 6.25$

Area of the circle = $\boxed{19.625 \text{ m}^2}$

(2 marks)

(b) Calculate the circumference of a circle of diameter 6 m.

$C = 2\pi r$ diameter = 6 m   radius = 3 m

$C = 2 \times 3.14 \times 3$ m

Circumference of the circle = $\boxed{18.84 \text{ m}}$

(2 marks)

(c) A circle has an area of 49 m². Calculate the diameter of the circle.

$A = \pi r^2$ making r the subject of the formula

$r^2 = \dfrac{A}{\pi}$

$r = \left(\dfrac{A}{\pi}\right)^{0.5}$

$r = \left(\dfrac{49}{3.14}\right)^{0.5}$ $r = 3.950328536 \, m$

d = 2r

$\boxed{\text{diameter = 7.90 m (to 2 decimal places)}}$

(2 marks)

Question 3

Consider the rectangle below:

(Diagram not drawn to scale)

5 cm

9 cm

(a) Calculate the perimeter of the rectangle.

Perimeter = $2 \times l + 2 \times w$

Perimeter = 2×9 cm + 2×5 cm

Perimeter = 18 cm + 10 cm

Perimeter = 28 cm

(1 mark)

(b) Calculate the area of the rectangle.

Area = length × width

Area = 9 cm × 5 cm

Area = 45 cm²

(1 mark)

(c) A square has an area of 144m². Determine the length of the side of the square.

$A = s^2$ s = $(A)^{0.5}$ s = $(144)^{0.5}$ **s = 12 m**

(1 mark)

Question 4

Calculate the area of each trapezium shown:

(Diagrams not drawn to scale)

(a)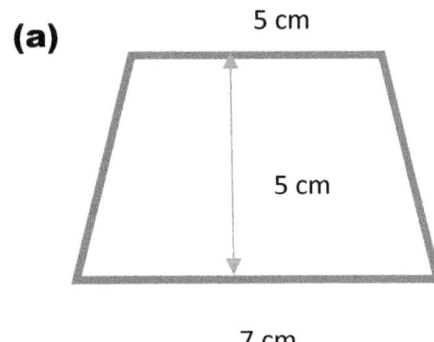

Area of a trapezium = $\frac{1}{2}(a + b) \times h$

Area of trapezium = $\frac{1}{2}(5cm + 7cm) \times 5cm$

Area of trapezium = $\frac{1}{2}(12cm) \times 5cm$

Area of trapezium = 30 cm²

(2 marks)

(b)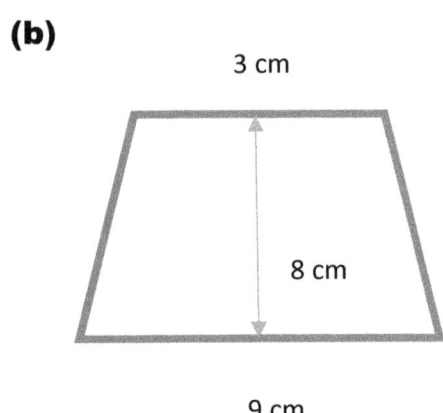

Area of a trapezium = $\frac{1}{2}(a + b) \times h$

Area of trapezium = $\frac{1}{2}(3 + 9) \times 8$

Area of trapezium = $\frac{1}{2}(12) \times 8$

Area of trapezium = 6×8

Area of trapezium = 48 cm²

(2 marks)

(c)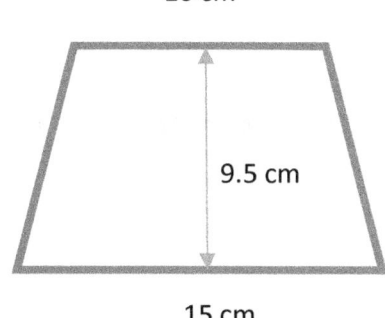

Area of a trapezium = $\frac{1}{2}(a + b) \times h$

Area of trapezium = $\frac{1}{2}(10cm + 15cm) \times 9.5$ cm

Area of trapezium = $\frac{1}{2}(25cm) \times 9.5$ cm

Area of trapezium = $12.5cm \times 9.5$ cm

Area of trapezium = 118.75 cm²

(2 marks)

(d)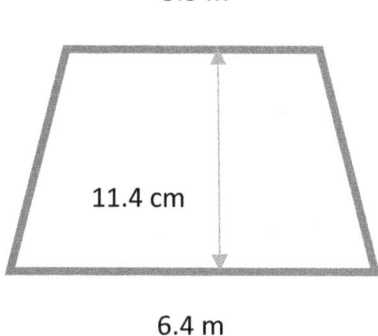

Area of a trapezium = $\frac{1}{2}(a + b) \times h$

Area of trapezium = $\frac{1}{2}(5.5 + 6.4) \times 0.114$

Area of trapezium = $0.5(11.9) \times 0.114$

Area of trapezium = 5.95×0.114

Area of trapezium = 0.6783 m²

(2 marks)

Question 5

Calculate the area of the triangle with the given dimensions.

(Diagram not drawn to scale)

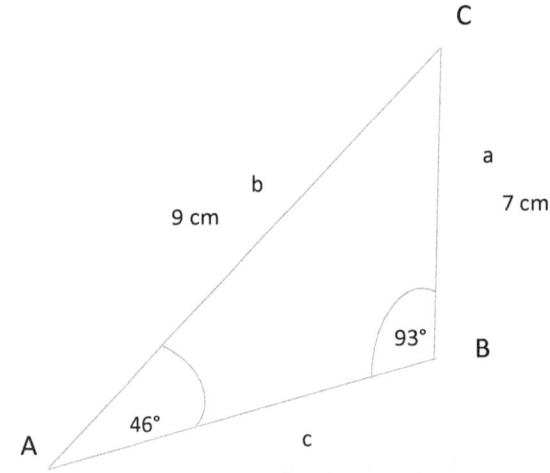

Area of a triangle = $\frac{1}{2}$ ab sin C

C = 180° - (46° + 93°)   (Internal angles in a triangle sum to 180°)

C = 41°

Area of triangle = $\frac{1}{2}$ (7cm) × (9 cm) sin (41°)

Area of triangle = $\frac{63}{2}$ sin (41°) = 31.5 sin (41°)

Area of triangle = 20.67 cm² (to 2 decimal places)

(2 marks)

Question 6

Calculate the area of the triangle with the given dimensions.

(Diagram not drawn to scale)

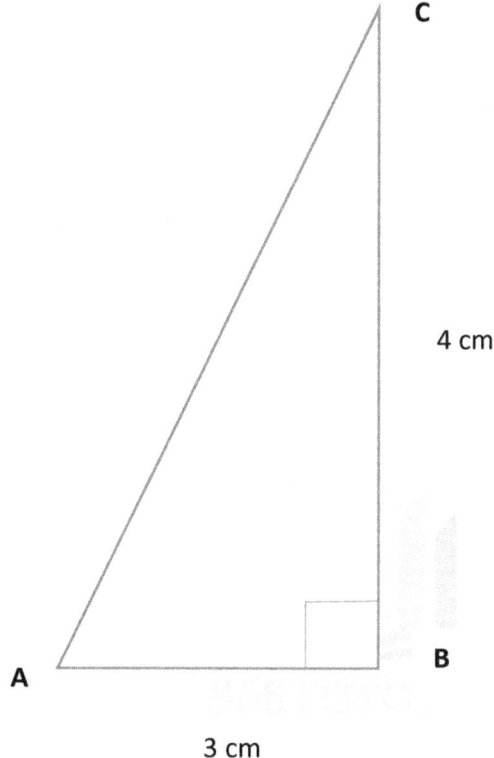

Area of a triangle = $\frac{1}{2}$ (b × h)

Area of triangle = $\frac{1}{2}$ (3cm × 4cm)

Area of triangle = 6 cm²

(2 marks)

Question 7

Calculate the area of the parallelograms with the given dimensions:

(Diagrams not drawn to scale)

(a)

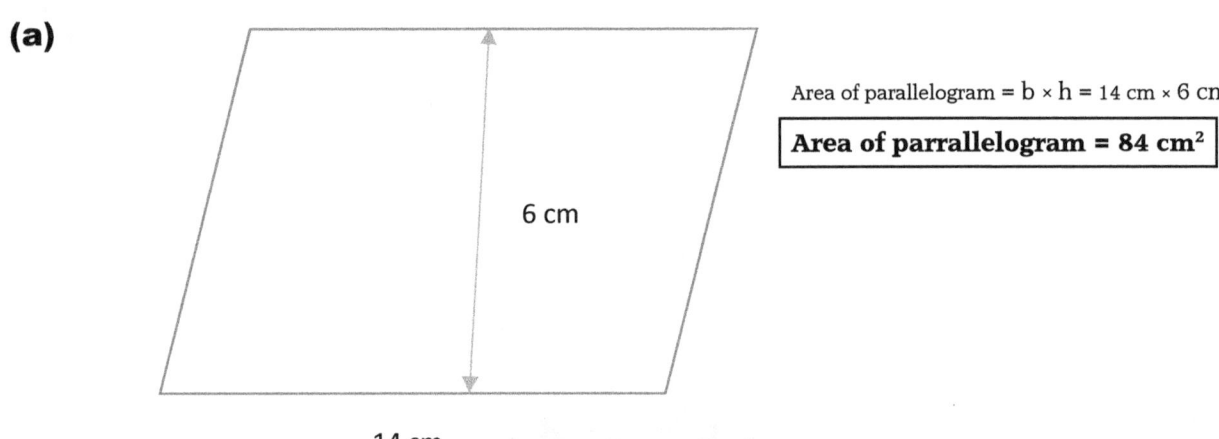

Area of parallelogram = b × h = 14 cm × 6 cm

Area of parrallelogram = 84 cm²

(2 marks)

(b)

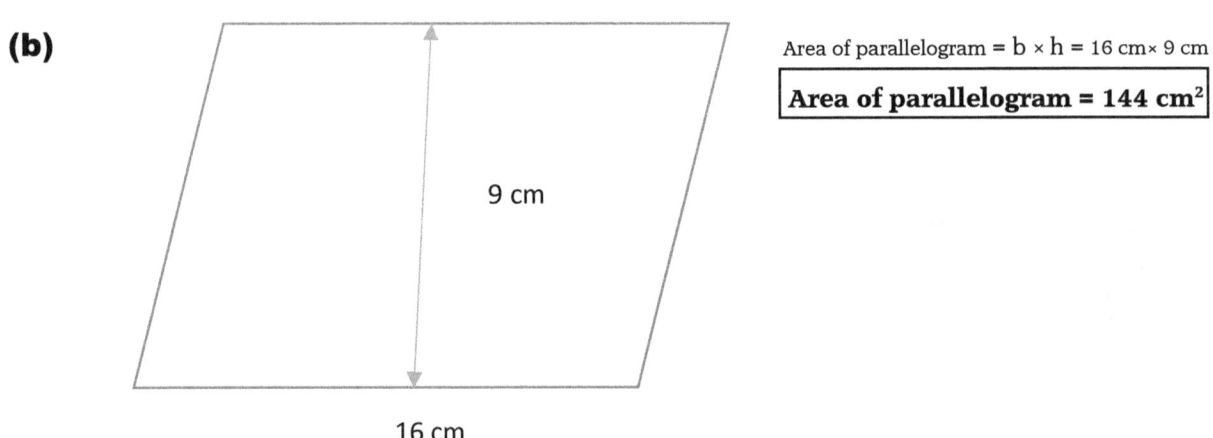

Area of parallelogram = b × h = 16 cm × 9 cm

Area of parrallelogram = 144 cm²

(2 marks)

Question 8

Consider a circle inside of a rectangle. The circle has a diameter of 40 m.

Use π = 3.14

(Diagram not drawn to scale)

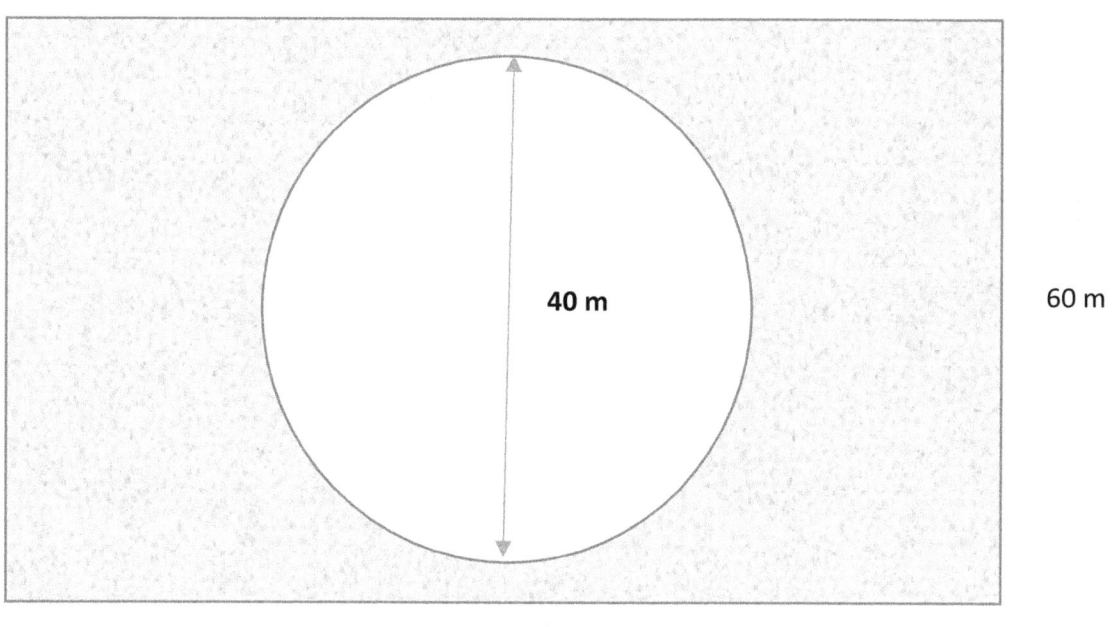

Calculate the area of the shaded region.

Area of the shaded region = Area of the rectangle − Area of the circle

Area of the rectangle = b × h = 110 m × 60 m = 6600 m²

Area of the circle = πr^2 radius = 20 m

Area of the circle = 3.14 × $(20)^2$ = 3.14 × 400 = 1256 m²

Area of the shaded region = 6600 m² − 1256 m²

Area of the shaded region = 5344 m²

(5 marks)

Question 9

Consider a triangle inside of a rectangle.

The triangle has a base of 6 m and a height of 8 m.

(Diagram not drawn to scale)

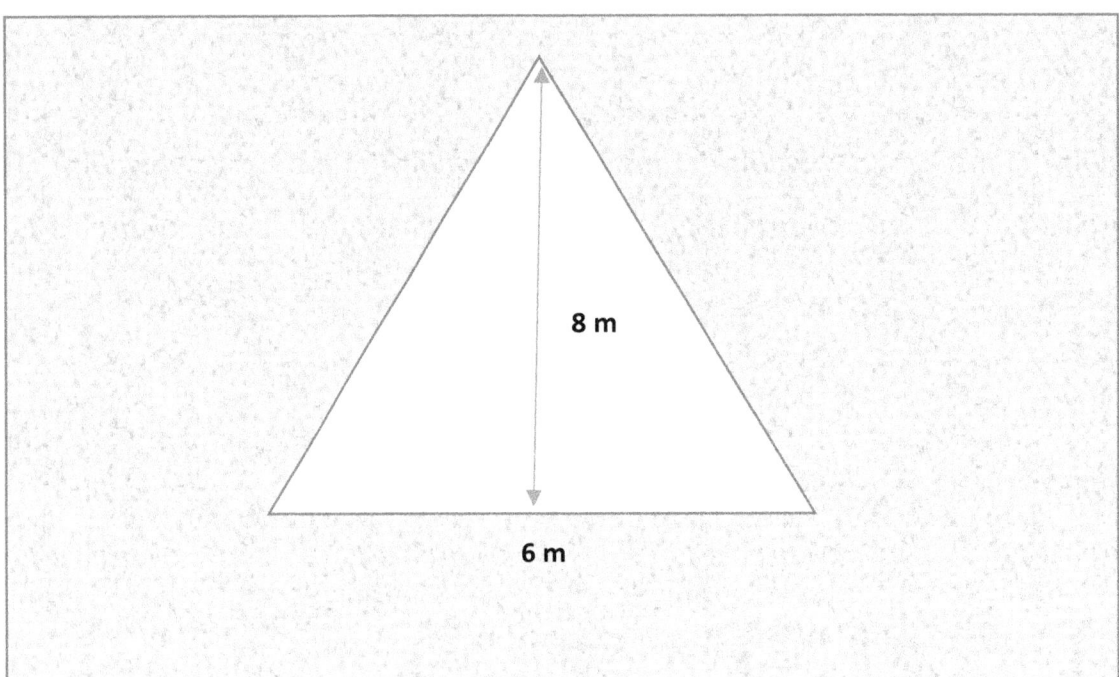

Calculate the area of the shaded region.

Area of the shaded region = Area of the rectangle – Area of the triangle

Area of the rectangle = b × h = 22 m × 20 m = 440 m²

Area of the triangle = $\frac{1}{2}$ (b × h) = $\frac{1}{2}$ (6 × 8) = 24 m²

Area of the shaded region = 440 m² - 24 m² = 416 m²

(5 marks)

Question 10

Consider the cylindrically shaped tank below. The radius of the circle that forms part of the tank is 7 m. The height of the tank is 10 m.

Use π = 3.14

(Diagram not drawn to scale)

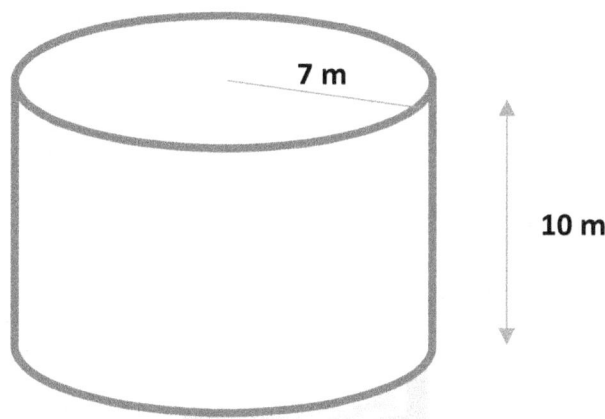

(a) Calculate the volume of the tank.

Volume of a cylindrical tank = $\pi r^2 h$

Volume of the tank = $(3.14) \times (7)^2 \times 10$ m

Volume of the tank = $(3.14) \times 49 \times 10$ m

Volume of the tank = 1538.6 m³

(5 marks)

Question 11

Consider the major circle sector with a sector angle of 270° and a radius of 20 cm.

Use π = 3.14

(Diagram not drawn to scale)

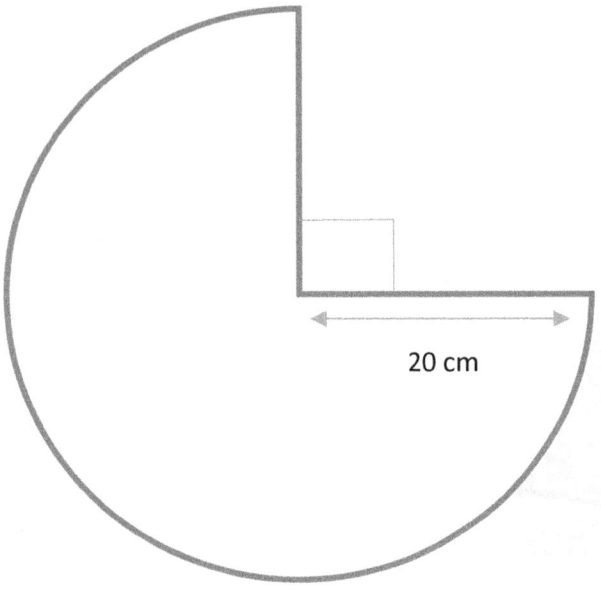

(a) Calculate the area of the major sector.

Area of a circle = πr^2

Area of the major sector = $\frac{270°}{360°} \times \pi r^2$

Area of the major sector = $\frac{3}{4} \times \pi r^2$ = 0.75 × (3.14) × (20)^2

Area of the major sector = 0.75 × (3.14) × 400 = $\boxed{942 \text{ cm}^2}$

(2 marks)

(b) Calculate the perimeter of the major sector.

Perimeter of the major sector = 20 cm + 20 cm + $\frac{3}{4}$ (2 πr)

Perimeter of the major sector = 40 cm + $\frac{3}{4}$ (2 × 3.14 × 20 cm) = $\boxed{134.2 \text{ cm}}$

(2 marks)

Question 12

Use π = 3.14

(a) Calculate the volume of a pyramid with a base of 20 m² and a height of 5 m.

Volume of a pyramid = Area of base × height

Volume of pyramid = 20 m² × 5 m

$\boxed{\text{Volume of pyramid = 100 m}^3}$

(2 marks)

(b) Calculate the volume of a sphere that has a radius of 200 m.

$Volume\ of\ a\ sphere = \frac{4}{3}\pi r^3$ radius = 200m

Volume of sphere = $\frac{4}{3} \times 3.14 \times (200)^3$ = $\frac{4}{3} \times 3.14 \times 8{,}000{,}000$

$\boxed{\text{Volume of sphere = 33,493,333.33 m}^3}$

(3 marks)

(c) Calculate the radius of a sphere that has a volume of 300 m³.

$Volume\ of\ a\ sphere = \frac{4}{3}\pi r^3$, making r the subject of the formula

$r = \left(\frac{0.75\ V}{\pi}\right)^{\frac{1}{3}}$ $r = \left(\frac{0.75 \times 300}{\pi}\right)^{\frac{1}{3}}$

$r = \left(\frac{225}{3.14}\right)^{\frac{1}{3}}$

$\boxed{r = 4.15 \text{ m (to 2 decimal places)}}$

(3 marks)

(d) Calculate the surface area of a sphere that has a radius of 20 m.

A (Surface area of a sphere) = $4\pi r^2$ radius = 20m

A = 4 × 3.14 × (20)² = $\boxed{\textbf{5,024 m}^2}$

(2 marks)

(e) A cone has a diameter of 30 cm and a vertical height of 64 cm. Calculate the volume of the cone.

V (Volume of a cone) = $\frac{\pi h r^2}{3}$ diameter = 30 cm, radius = 15 cm

$$V = \frac{3.14 \times 64 cm \times (15)^2}{3}$$

V= 15,072 cm³

(2 marks)

(f) A cone has a volume of 320 cm³. The cone has a height of 40 cm. Calculate the radius of the cone.

V (Volume of a cone) = $\frac{\pi h r^2}{3}$

Making r the subject of the formula

3v = $\pi h r^2$

r = $\left(\frac{3V}{\pi \times h}\right)^{\frac{1}{2}}$ r = $\left(\frac{3 \times 320}{3.14 \times 40}\right)^{\frac{1}{2}}$ r = $\left(\frac{960}{125.6}\right)^{\frac{1}{2}}$ radius = 2.76 cm (to 2 decimal places)

(2 marks)

END OF WORKSHEET

868

868TUTORS

Preparation for

High School Mathematics

Quadratic Equations

Solutions

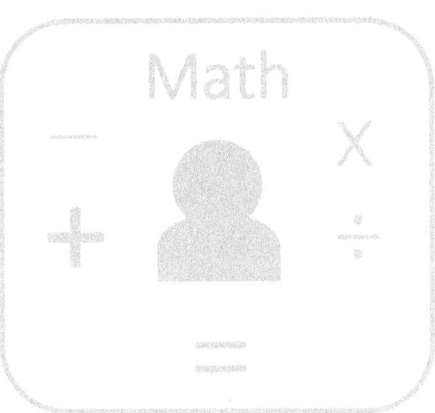

Instructions and Tips:

- ✓ You have 60 minutes to complete this worksheet
- ✓ This worksheet consists of 7 questions
- ✓ Write answers in the spaces provided
- ✓ Show all working

Student Name: _____

Student ID: _____

Date: __/__/____

Total Score:

Highest Score:

Tutor's Comments:

Access more free worksheets at www.868tutors.com

Question 1

Solve the following equations:

(a) $x^2 + 3x + 2 = 0$

Factorisation possible

$(x + 2)(x + 1) = 0$

$x + 2 = 0$ or $x + 1 = 0$

$\boxed{x = -2 \text{ or } x = -1}$

(b) $x^2 + 5x + 4 = 0$

Factorisation possible

$(x + 4)(x + 1) = 0$

$x + 4 = 0$ or $x + 1 = 0$

$\boxed{x = -4 \text{ or } x = -1}$

(c) $x^2 + 6x + 8 = 0$

Factorisation possible

$(x + 4)(x + 2) = 0$

$x + 4 = 0$ or $x + 2 = 0$

$\boxed{x = -4 \text{ or } x = -2}$

(6 marks)

Question 2

Solve the following equations:

(a) $2x^2 + 2x - 4 = 0$

$\div$ throughout by 2
$x^2 + 1x - 2 = 0$
Factorisation possible
$(x + 2)(x - 1) = 0$
$x + 2 = 0$ or $x - 1 = 0$

$\boxed{x = -2 \text{ or } x = 1}$

(b) $4x^2 - 2x - 2 = 0$

$\div$ throughout by 2
$2x^2 - 1x - 1 = 0$
Factorisation possible
$(2x + 1)(x - 1) = 0$
$2x + 1 = 0$ or $x - 1 = 0$
$2x = -1$ or $x = 1$

$\boxed{x = \frac{-1}{2} \text{ or } x = 1}$

(c) $x^2 + 4x - 32 = 0$

Factorisation possible
$(x + 8)(x - 4) = 0$

$\boxed{x = -8 \text{ or } x = 4}$

(9 marks)

Question 3

Solve the following equations (to 2 decimal places):

(a) $x^2 + x - 10 = 0 \qquad a = 1 \quad b = 1 \quad c = -10$

Applying the quadratic formula $\quad x = \dfrac{-b \pm \sqrt{b^2 - 4ac}}{2a}$

$x = \dfrac{-b + \sqrt{b^2 - 4ac}}{2a} \quad or \quad x = \dfrac{-b - \sqrt{b^2 - 4ac}}{2a}$

$x = \dfrac{-1 + \sqrt{(1)^2 - 4(1)(-10)}}{2(1)}$

$x = \dfrac{-1 + \sqrt{(1)^2 + 40}}{2}$

$x = \dfrac{-1 + \sqrt{41}}{2} \quad or \quad x = \dfrac{-1 - \sqrt{41}}{2}$

x = 2.70 or x = -3.70

(b) $x^2 + 9x + 4 = 0 \qquad a = 1 \quad b = 9 \quad c = 4$

Applying the quadratic formula $\quad x = \dfrac{-b \pm \sqrt{b^2 - 4ac}}{2a}$

$x = \dfrac{-b + \sqrt{b^2 - 4ac}}{2a} \quad or \quad x = \dfrac{-b - \sqrt{b^2 - 4ac}}{2a}$

$x = \dfrac{-9 + \sqrt{(9)^2 - 4(1)(4)}}{2(1)}$

$x = \dfrac{-9 + \sqrt{81 - 16}}{2}$

$x = \dfrac{-9 + \sqrt{65}}{2} \quad or \quad x = \dfrac{-9 - \sqrt{65}}{2}$

x = - 0.47 or x = - 8.53

(10 marks)

Question 4

Solve the following equations (to 2 decimal places):

(a) $-x^2 + 7x - 5 = 0 \qquad a = -1 \ \ b = 7 \ c = -5$

Applying the quadratic formula $\quad x = \dfrac{-b \pm \sqrt{b^2 - 4ac}}{2a}$

$x = \dfrac{-b + \sqrt{b^2 - 4ac}}{2a} \quad$ or $\quad x = \dfrac{-b - \sqrt{b^2 - 4ac}}{2a}$

$x = \dfrac{-7 + \sqrt{(7)^2 - 4(-1)(-5)}}{2(-1)}$

$x = \dfrac{-7 + \sqrt{49 - 20}}{-2}$

$x = \dfrac{-7 + \sqrt{29}}{-2} \quad$ or $\quad x = \dfrac{-7 - \sqrt{29}}{-2}$

$\boxed{x = 0.81 \text{ or } x = 6.19}$

(b) $-x^2 - 5x + 11 = 0 \qquad a = -1 \ \ b = -5 \ \ c = 11$

Applying the quadratic formula $\quad x = \dfrac{-b \pm \sqrt{b^2 - 4ac}}{2a}$

$x = \dfrac{-b + \sqrt{b^2 - 4ac}}{2a} \quad$ or $\quad x = \dfrac{-b - \sqrt{b^2 - 4ac}}{2a}$

$x = \dfrac{-(-5) + \sqrt{(-5)^2 - 4(-1)(11)}}{2(-1)}$

$x = \dfrac{5 + \sqrt{25 - -44}}{-2}$

$x = \dfrac{5 + \sqrt{69}}{-2} \quad$ or $\quad x = \dfrac{5 - \sqrt{69}}{-2}$

$\boxed{x = -6.65 \text{ or } x = 1.65}$

(10 marks)

Question 5

Solve the following equations (to 2 decimal places):

(a) $2x^2 + 3x - 1 = 0 \quad a = 2 \quad b = 3 \quad c = -1$

Applying the quadratic formula $\quad x = \dfrac{-b \pm \sqrt{b^2 - 4ac}}{2a}$

$x = \dfrac{-b + \sqrt{b^2 - 4ac}}{2a}$ or $\quad x = \dfrac{-b - \sqrt{b^2 - 4ac}}{2a}$

$x = \dfrac{-3 + \sqrt{(3)^2 - 4(2)(-1)}}{2(2)}$

$x = \dfrac{-3 + \sqrt{9 - -8}}{4}$

$x = \dfrac{-3 + \sqrt{17}}{4}$ or $x = \dfrac{-3 - \sqrt{17}}{4}$

$\boxed{x = 0.28 \text{ or } x = -1.78}$

(b) $3x^2 + 8x - 1 = 0 \quad a = 3 \quad b = 8 \quad c = -1$

Applying the quadratic formula $\quad x = \dfrac{-b \pm \sqrt{b^2 - 4ac}}{2a}$

$x = \dfrac{-b + \sqrt{b^2 - 4ac}}{2a}$ or $\quad x = \dfrac{-b - \sqrt{b^2 - 4ac}}{2a}$

$x = \dfrac{-8 + \sqrt{(8)^2 - 4(3)(-1)}}{2(3)}$

$x = \dfrac{-8 + \sqrt{64 - -12}}{6}$

$x = \dfrac{-8 + \sqrt{76}}{6}$ or $\dfrac{-8 - \sqrt{76}}{6}$

$\boxed{x = 0.12 \text{ or } x = -2.79}$

(10 marks)

Question 6

Solve the following equations (to 2 decimal places):

(a) $5a^2 + 12a - 1 = 0$ $d = 5$ $b = 12$ $c = -1$

Applying the quadratic formula $x = \frac{-b \pm \sqrt{b^2 - 4dc}}{2d}$ ← Quadratic formula modified for question

$a = \frac{-b + \sqrt{b^2 - 4dc}}{2d}$ or $x = \frac{-b - \sqrt{b^2 - 4dc}}{2d}$

$a = \frac{-12 + \sqrt{(12)^2 - 4(5)(-1)}}{2(5)}$

$a = \frac{-12 + \sqrt{144 - -20}}{10}$

$a = \frac{-12 + \sqrt{164}}{10}$ or $a = \frac{-12 - \sqrt{164}}{10}$

$\boxed{a = 0.08 \text{ or } a = -2.48}$

(b) $3b^2 + 4b + 1 = 0$ $a = 3$ $b = 4$ $c = 1$

Factorisation possible

$3b^2 + 4b + 1 = 0$

$3b^2 + 3b + b + 1 = 0$

$3b(b+1) + 1(b+1) = 0$

$(3b+1)(b+1) = 0$

$3b + 1 = 0$ or $b + 1 = 0$

$3b + 1 = 0$

$3b = -1$ or $b + 1 = 0$

$\boxed{b = \frac{-1}{3} \text{ or } b = -1}$

(10 marks)

Question 7

Consider the rectangular section of land for agriculture:

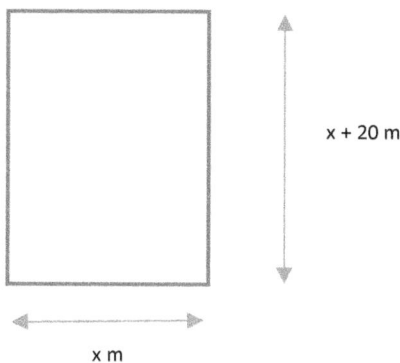

The width of the land is **x** m and the length of the land is **x + 20**m. The area of the rectangular section is 1400 m². Calculate the width of the land (to 2 decimal places).

Area of rectangular section = width × length

Area of rectangular section = $x \times (x + 20)$

$1400 = x^2 + 20x$

$x^2 + 20x = 1400$

$x^2 + 20x - 1400 = 0$ (Now we need to solve this quadratic equation)

a = 1   b = 20   c = -1400

Applying the quadratic formula $x = \frac{-b \pm \sqrt{b^2-4ac}}{2a}$

$x = \frac{-b + \sqrt{b^2-4ac}}{2a}$ or $x = \frac{-b - \sqrt{b^2-4ac}}{2a}$

$x = \frac{-20 + \sqrt{(20)^2 - 4(1)(-1400)}}{2(1)}$

$x = \frac{-20 + \sqrt{400 - -5600}}{2}$

$x = \frac{-20 + \sqrt{6000}}{2}$ or $\frac{-20 - \sqrt{6000}}{2}$

x = 28.73 or -48.73   (Use the positive value, since length must have a positive value)

The width of the land is 28.73 m.   (8 marks)

END OF WORKSHEET

868 TUTORS

Preparation for

High School Mathematics

Quadratic Graphs

Solutions

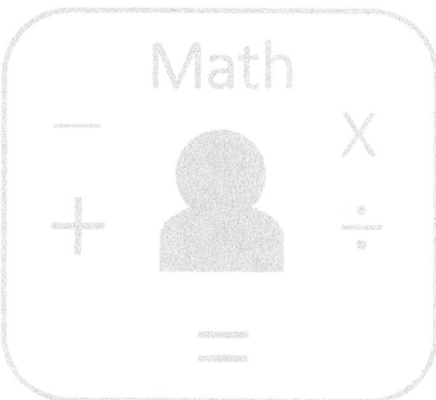

Instructions and Tips:

- You have 90 minutes to complete this worksheet
- This worksheet consists of 7 questions
- Write answers in the spaces provided
- All working must be clearly shown
- Label Graphs properly
- Draw Smooth Curves

Student Name: _____

Student ID: _____

Date: __ / __ / ____

Total Score:

Highest Score:

Tutor's Comments:

Access more free worksheets at www.868tutors.com

Question 1

Consider the quadratic function: $y = x^2 + 4x + 4$.

x	-5	-4	-3	-2	-1	0	1
y	9	4	1	0	1	4	9

(a) Complete the table above for: $y = x^2 + 4x + 4$.

$y = x^2 + 4x + 4$
when $x = -5$
$y = (-5)^2 + 4(-5) + 4$
$y = 25 - 20 + 4$
$y = 5 + 4$
y = 9

$y = x^2 + 4x + 4$
when $x = -4$
$y = (-4)^2 + 4(-4) + 4$
$y = 16 - 16 + 4$
$y = 0 + 4$
y = 4

$y = x^2 + 4x + 4$
when $x = -3$
$y = (-3)^2 + 4(-3) + 4$
$y = 9 - 12 + 4$
$y = -3 + 4$
y = 1

$y = x^2 + 4x + 4$
when $x = -1$
$y = (-1)^2 + 4(-1) + 4$
$y = 1 - 4 + 4$
$y = -3 + 4$
y = 1

$y = x^2 + 4x + 4$
when $x = 0$
$y = (0)^2 + 4(0) + 4$
$y = 0 + 0 + 4$
$y = 0 + 4$
y = 4

(4 marks)

(b) On the graph paper on the next page, draw the graph of

$y = x^2 + 4x + 4$ using the table above. Use a scale of 2cm = 1 unit on the x–axis and 1 cm = 1 unit on the y–axis.

(6 marks)

Question 1 : $y = x^2 + 4x + 4$

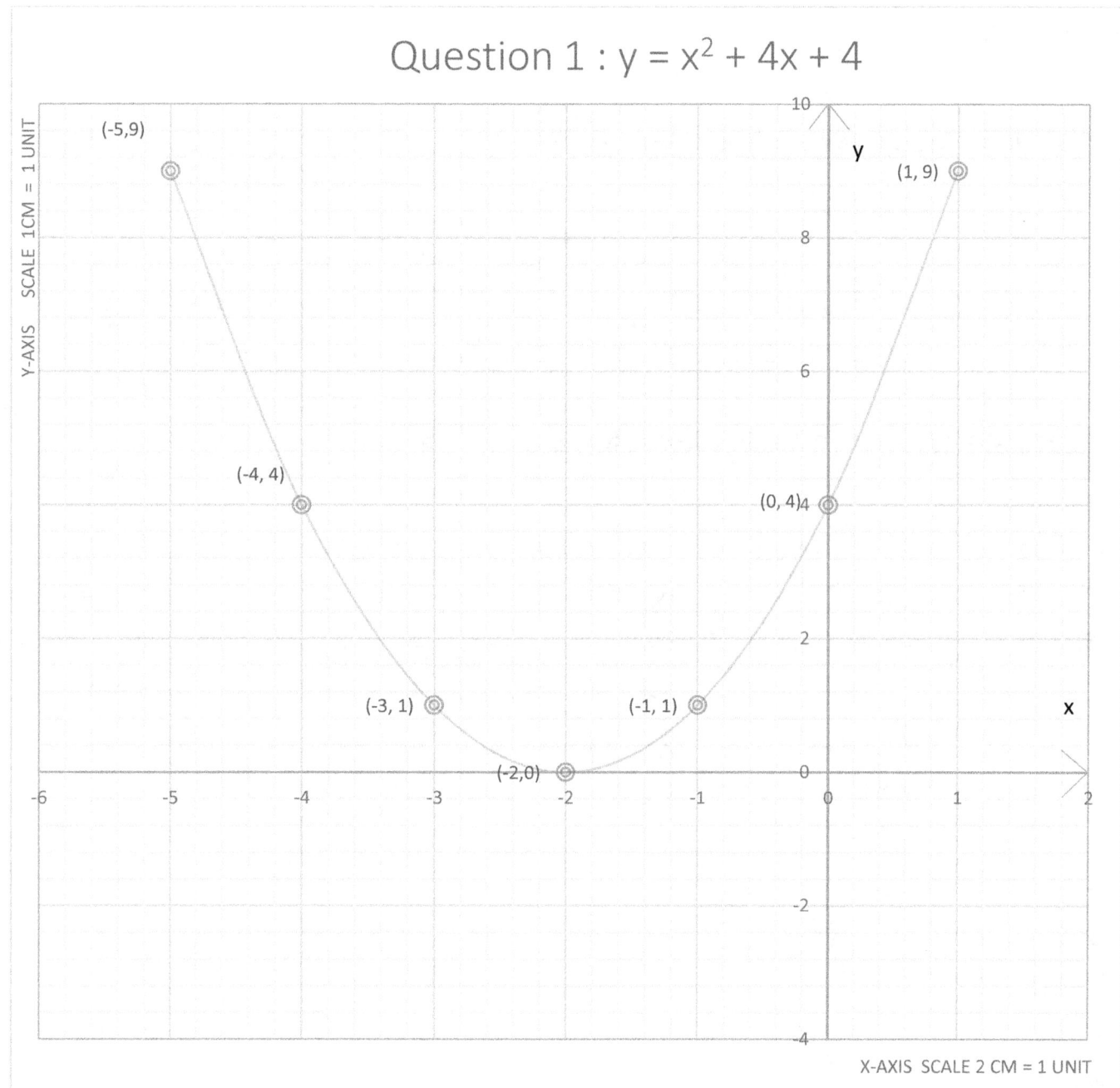

Question 2

Consider the quadratic function: $y = x^2 + 3x + 2$.

x	-5	-4	-3	-2	-1	0	1	2
y	12	6	2	0	0	2	6	12

(a) Complete the table above for: $y = x^2 + 3x + 2$.

$y = x^2 + 3x + 2$
when x = -5
$y = (-5)^2 + 3(-5) + 2$
$y = 25 - 15 + 2$
$y = 10 + 2$
y = 12

$y = x^2 + 3x + 2$
when x = -4
$y = (-4)^2 + 3(-4) + 2$
$y = 16 - 12 + 2$
$y = 4 + 2$
y = 6

$y = x^2 + 3x + 2$
when x = -2
$y = (-2)^2 + 3(-2) + 2$
$y = 4 - 6 + 2$
$y = -2 + 2$
y = 0

$y = x^2 + 3x + 2$
when x = -1
$y = (-1)^2 + 3(-1) + 2$
$y = 1 - 3 + 2$
$y = -2 + 2$
y = 0

$y = x^2 + 3x + 2$
when x = 0
$y = (0)^2 + 3(0) + 2$
$y = 0 + 0 + 2$
y = 2

$y = x^2 + 3x + 2$
when x = 1
$y = (1)^2 + 3(1) + 2$
$y = 1 + 3 + 2$
$y = 4 + 2$
y = 6

(4 marks)

(b) On the graph paper on the next page, draw the graph of

$y = x^2 + 3x + 2$ using the table above. Use an appropriate scale.

(6 marks)

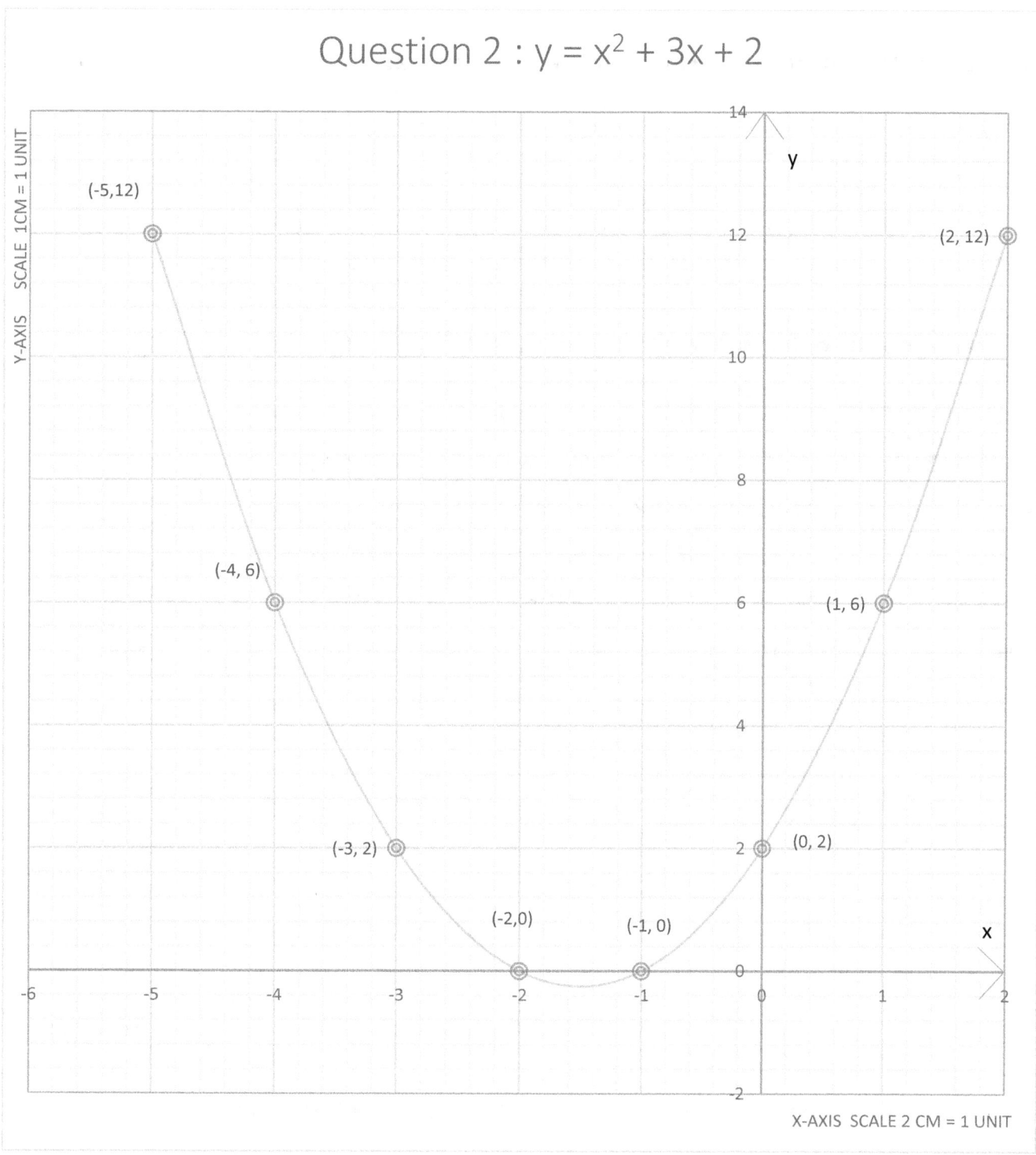

Question 3

Consider the quadratic function: $y = -x^2 + x + 2$.

x	-3	-2	-1	0	1	2	3	4
y	-10	-4	0	2	2	0	-4	-10

(a) Complete the table above for: $y = -x^2 + x + 2$.

$y = -x^2 + x + 2$
when $x = -2$
$y = -(-2)^2 + -2 + 2$
$y = -4 - 2 + 2$
$y = -6 + 2$
y = -4

$y = -x^2 + x + 2$
when $x = -1$
$y = -(-1)^2 + -1 + 2$
$y = -1 - 1 + 2$
$y = -2 + 2$
y = 0

$y = -x^2 + x + 2$
when $x = 0$
$y = -(0)^2 + 0 + 2$
$y = 0 + 0 + 2$
$y = 0 + 2$
y = 2

$y = -x^2 + x + 2$
when $x = 1$
$y = -(1)^2 + 1 + 2$
$y = -1 + 1 + 2$
$y = 0 + 2$
y = 2

$y = -x^2 + x + 2$
when $x = 3$
$y = -(3)^2 + 3 + 2$
$y = -9 + 5$
y = -4

$y = -x^2 + x + 2$
when $x = 4$
$y = -(4)^2 + 4 + 2$
$y = -16 + 6$
y = -10

(4 marks)

(b) On the graph paper on the next page, draw the graph of

$y = -x^2 + x + 2$ using the table above. Use an appropriate scale.

(6 marks)

Question 3 : $y = -x^2 + x + 2$

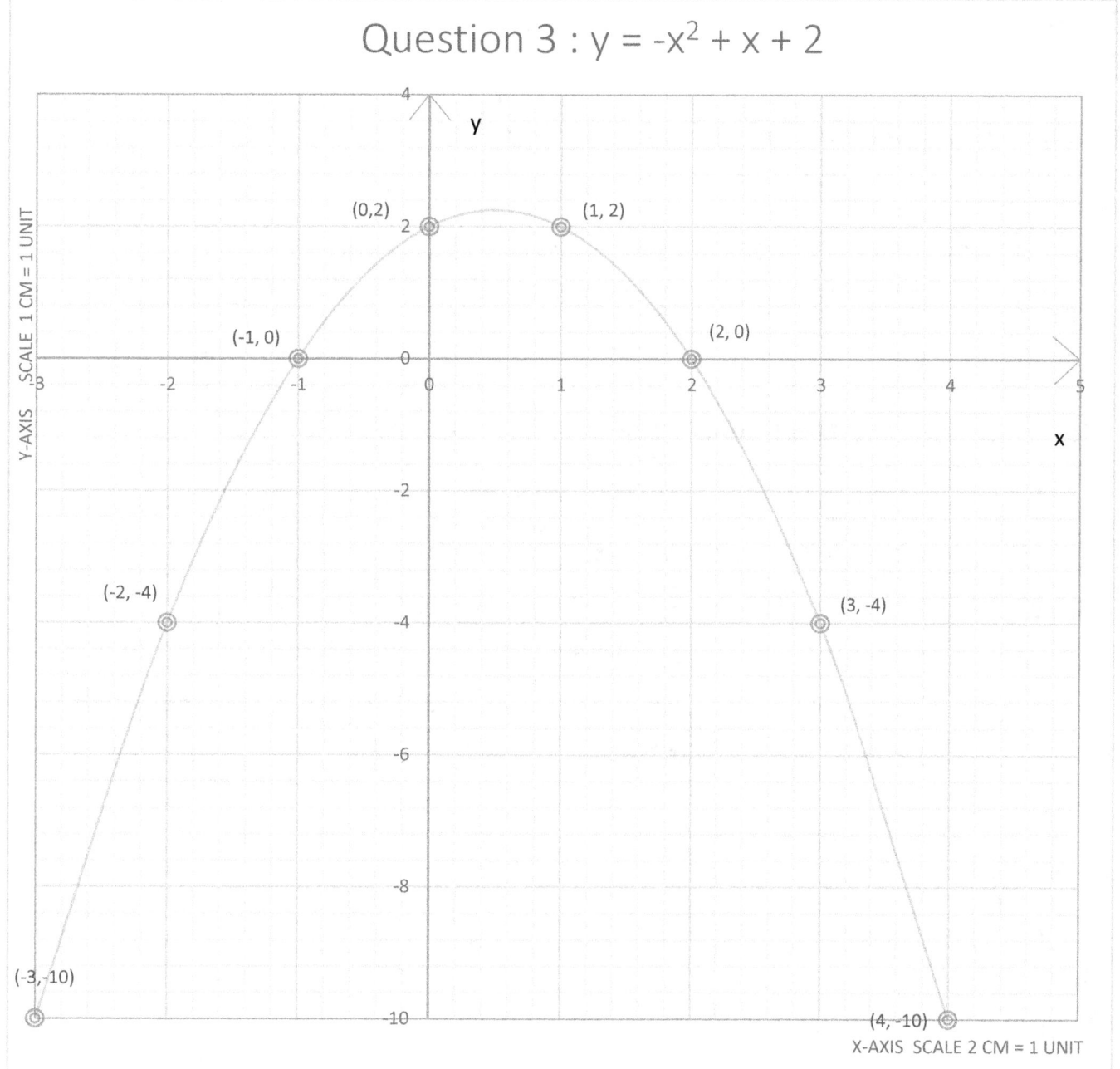

Question 4

Consider the quadratic function: $y = -x^2 + 2x + 3$.

x	-2	-1	0	1	2	3	4	5
y	-5	0	3	4	3	0	-5	-12

(a) Complete the table above for: $y = -x^2 + 2x + 3$.

$y = -x^2 + 2x + 3$
when $x = -1$
$y = -(1)^2 + 2(-1) + 3$
$y = -1 - 2 + 3$
$y = -3 + 3$
y = 0

$y = -x^2 + 2x + 3$
when $x = 0$
$y = -(0)^2 + 2(0) + 3$
$y = 0 + 0 + 3$
$y = 3$
y = 3

$y = -x^2 + 2x + 3$
when $x = 2$
$y = -(2)^2 + 2(2) + 3$
$y = -4 + 4 + 3$
$y = 3$
y = 3

$y = -x^2 + 2x + 3$
when $x = 3$
$y = -(3)^2 + 2(3) + 3$
$y = -9 + 6 + 3$
$y = -9 + 9$
y = 0

$y = -x^2 + 2x + 3$
when $x = 4$
$y = -(4)^2 + 2(4) + 3$
$y = -16 + 8 + 3$
$y = -8 + 3$
y = -5

$y = -x^2 + 2x + 3$
when $x = 5$
$y = -(5)^2 + 2(5) + 3$
$y = -25 + 10 + 3$
$y = -25 + 13$
y = -12

(5 marks)

(b) On the graph paper on the next page, draw the graph of

$y = -x^2 + 2x + 3$ using the table above. Use an appropriate scale.

(6 marks)

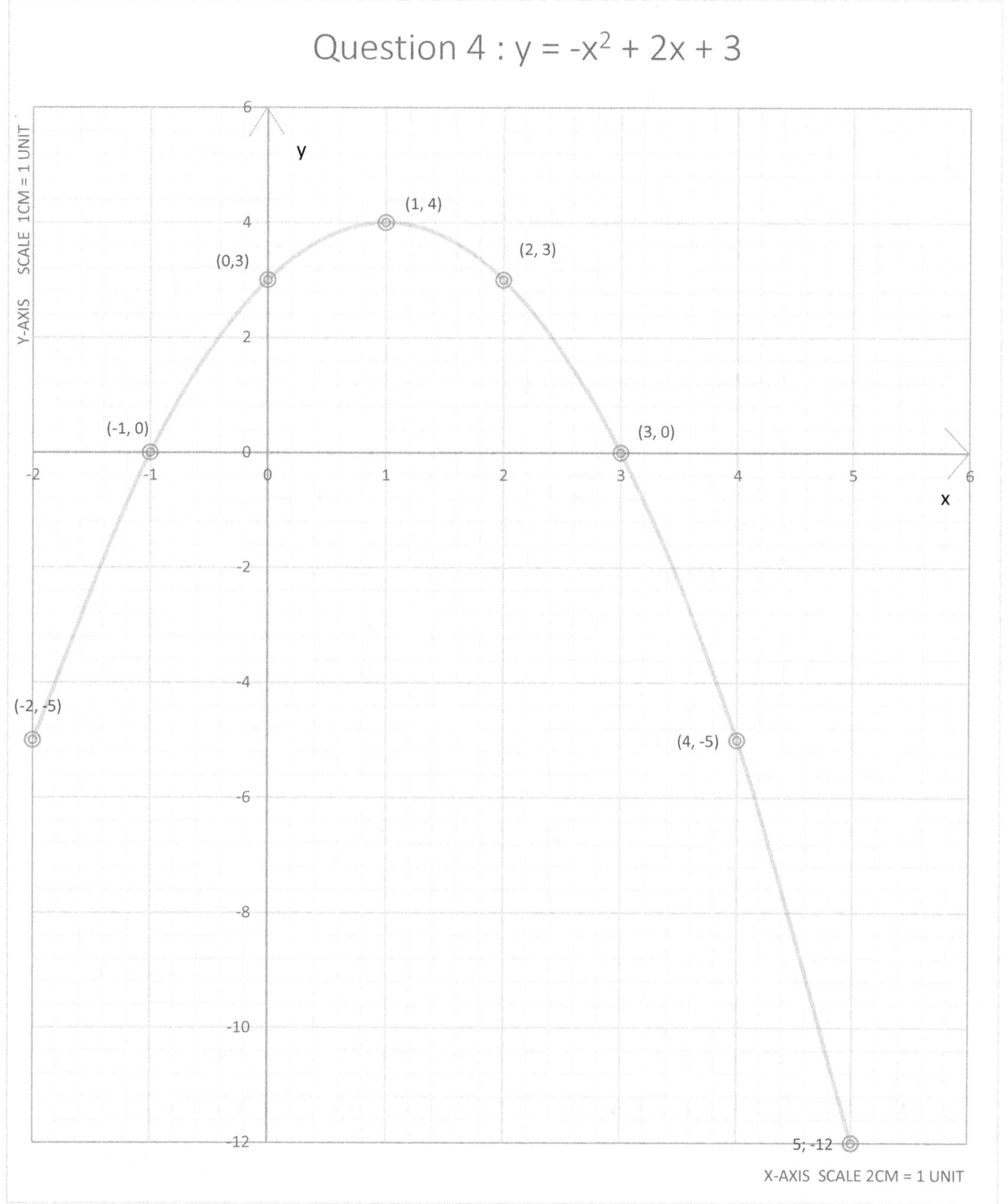

Question 5

Consider the quadratic function: $y = x^2 + 5x + 6$.

x	-6	-5	-4	-3	-2	-1	0	1
y	12	6	2	0	0	2	6	12

(a) Complete the table above for: $y = x^2 + 5x + 6$.

$y = x^2 + 5x + 6$
when $x = -6$
$y = (-6)^2 + 5(-6) + 6$
$y = 36 - 30 + 6$
$y = 6 + 6$
y = 12

$y = x^2 + 5x + 6$
when $x = -5$
$y = (-5)^2 + 5(-5) + 6$
$y = 25 - 25 + 6$
$y = 0 + 6$
y = 6

$y = x^2 + 5x + 6$
when $x = -4$
$y = (-4)^2 + 5(-4) + 6$
$y = 16 - 20 + 6$
$y = -4 + 6$
y = 2

$y = x^2 + 5x + 6$
when $x = -3$
$y = (-3)^2 + 5(-3) + 6$
$y = 9 - 15 + 6$
$y = -6 + 6$
y = 0

$y = x^2 + 5x + 6$
when $x = -2$
$y = (-2)^2 + 5(-2) + 6$
$y = 4 - 10 + 6$
$y = -6 + 6$
y = 0

$y = x^2 + 5x + 6$
when $x = -1$
$y = (-1)^2 + 5(-1) + 6$
$y = 1 - 5 + 6$
$y = -4 + 6$
y = 2

$y = x^2 + 5x + 6$
when $x = 0$
$y = (0)^2 + 5(0) + 6$
$y = 0 + 6$
y = 6

$y = x^2 + 5x + 6$
when $x = 1$
$y = (1)^2 + 5(1) + 6$
$y = 1 + 5 + 6$
y = 12

(7 marks)

(b) On the graph paper on the next page, draw the graph of $y = x^2 + 5x + 6$ using the table above. Use an appropriate scale.

(6 marks)

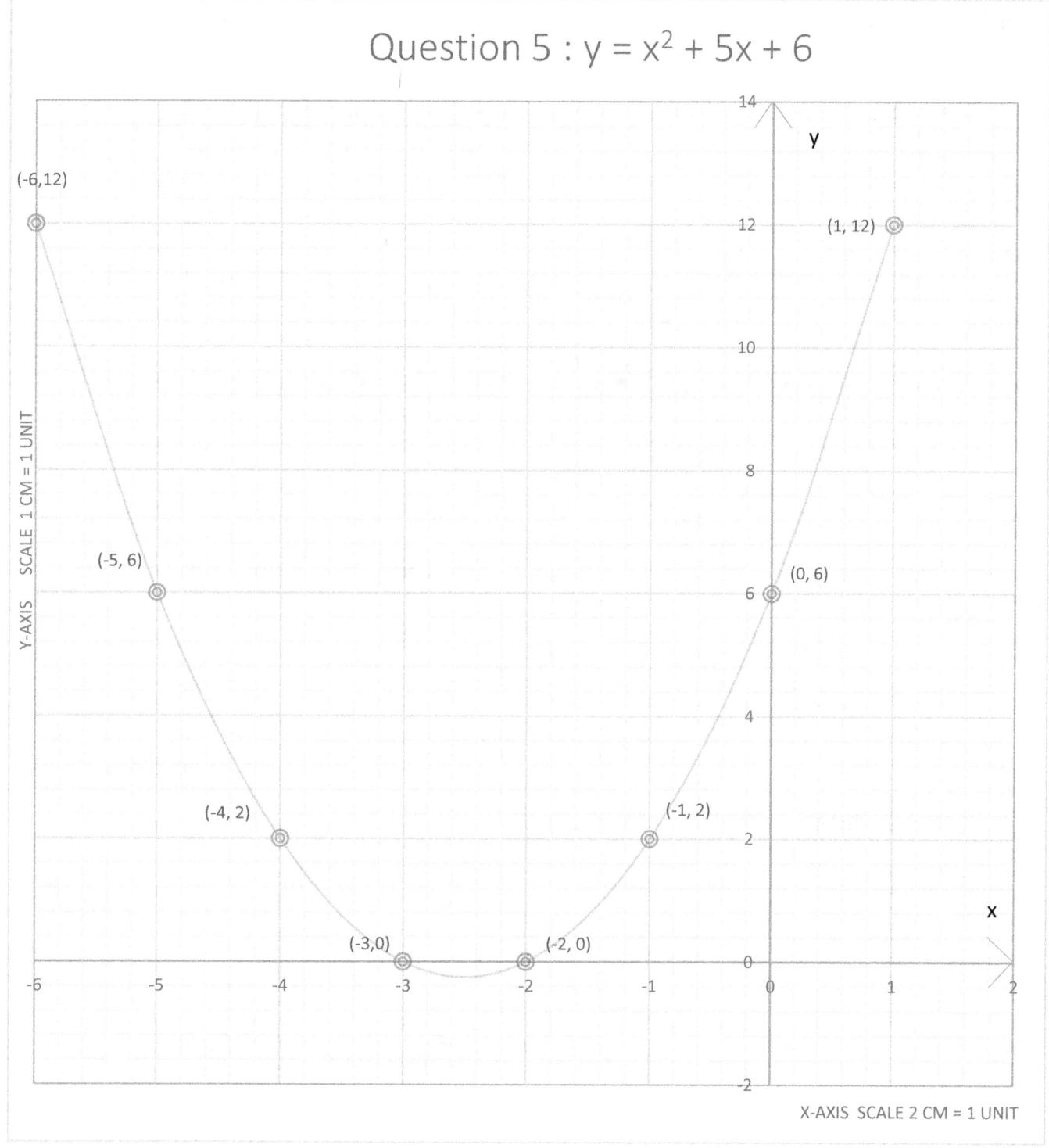

Complete the following statements using information from your graph.

(c) The values of x for which $x^2 + 5x + 6 = 0$ occur are at ____-3____

and ____-2____

(2 marks)

(d) The equation of the axis of symmetry of the graph $y = x^2 + 5x + 6$ is

____$x = -2\frac{1}{2}$____

The axis of symmetry is a vertical line that runs through the turning point of the parabola.

(2 marks)

(e) The minimum value of $x^2 + 5x + 6$ is ____$-\frac{1}{4}$____

(1 mark)

(f) The name of the graph of a quadratic function is a ____parabola____

(1 mark)

Question 6

Consider the quadratic function: y = x² - 5x + 4.

x	-1	0	1	2	3	4	5	6
y	10	4	0	-2	-2	0	4	10

(a) Complete the table above for: y = x² - 5x + 4.

$y = x^2 - 5x + 4$
when x = -1
$y = (-1)^2 - 5(-1) + 4$
$y = 1 + 5 + 4$
$y = 6 + 4$
y = 10

$y = x^2 - 5x + 4$
when x = 0
$y = (0)^2 - 5(0) + 4$
$y = 0 + 0 + 4$
$y = 4$
y = 4

$y = x^2 - 5x + 4$
when x = 1
$y = (-1)^2 - 5(1) + 4$
$y = 1 - 5 + 4$
$y = -4 + 4$
y = 0

$y = x^2 - 5x + 4$
when x = 2
$y = (2)^2 - 5(2) + 4$
$y = 4 - 10 + 4$
$y = -6 + 4$
y = -2

$y = x^2 - 5x + 4$
when x = 3
$y = (3)^2 - 5(3) + 4$
$y = 9 - 15 + 4$
$y = -6 + 4$
y = -2

$y = x^2 - 5x + 4$
when x = 4
$y = (4)^2 - 5(4) + 4$
$y = 16 - 20 + 4$
$y = -4 + 4$
y = 0

$y = x^2 - 5x + 4$
when x = 5
$y = (5)^2 - 5(5) + 4$
$y = 25 - 25 + 4$
$y = 0 + 4$
y = 4

$y = x^2 - 5x + 4$
when x = 6
$y = (6)^2 - 5(6) + 4$
$y = 36 - 30 + 4$
$y = 6 + 4$
y = 10

(10 marks)

(b) On the graph paper on the next page, draw the graph of

y = x² - 5x + 4 using the table above. Use an appropriate scale.

(6 marks)

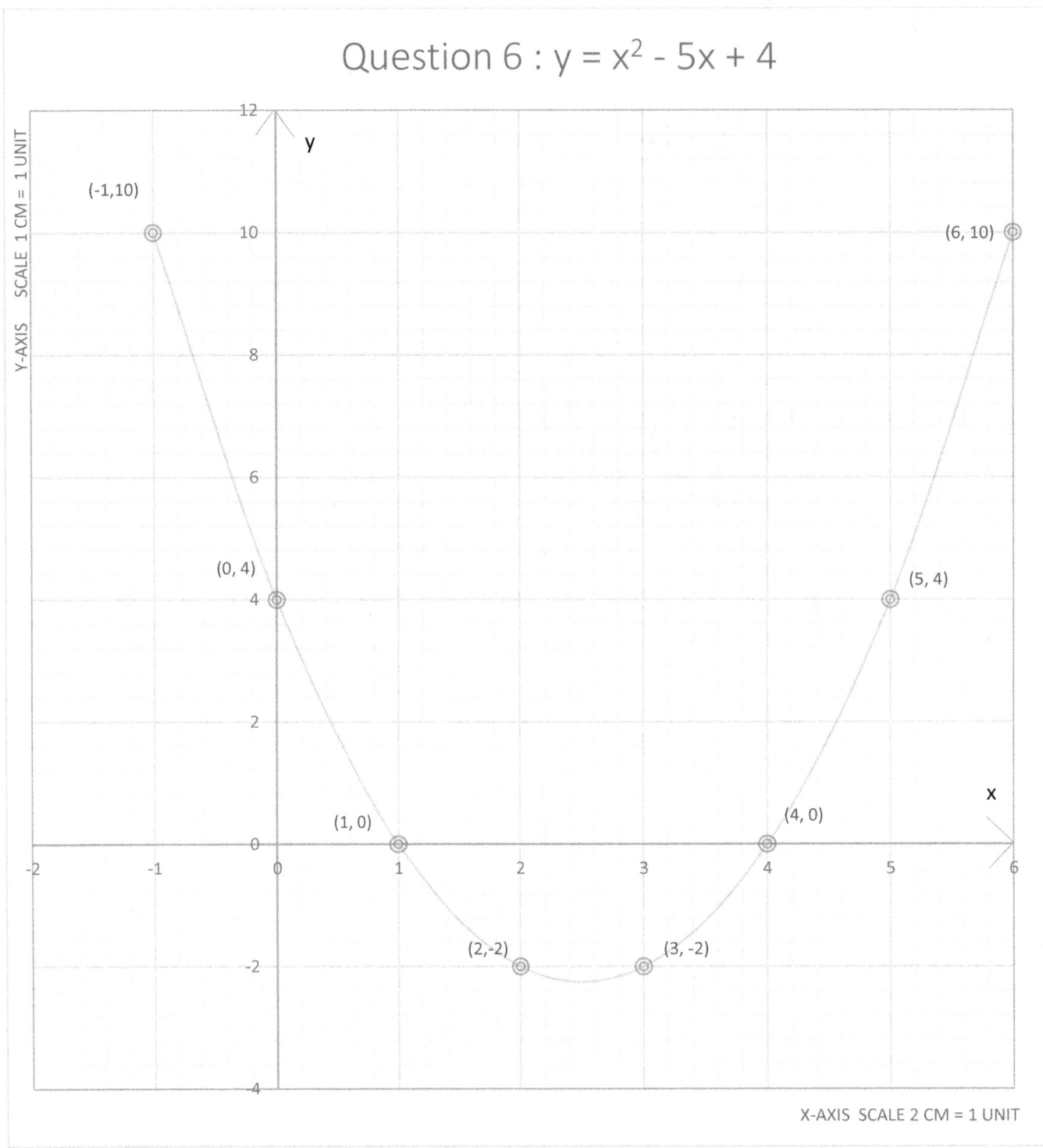

Complete the following statements using information from your graph.

(c) The values of x for which $x^2 - 5x + 4 = 0$ occur are at ___1___ and ___4___

(2 marks)

(d) The equation of the axis of symmetry of the graph $y = x^2 - 5x + 4$ is ___$x = 2\frac{1}{2}$___

(2 marks)

(e) The minimum value of $x^2 - 5x + 4$ is ___$-2\frac{1}{4}$___

(1 mark)

(f) The name of the graph of a quadratic function is a ___parabola___

(1 mark)

Question 7

Consider the quadratic function: $y = -x^2 + x + 2$.

x	-3	-2	-1	0	1	2	3	4
y	-10	-4	0	2	2	0	-4	-10

(a) Complete the table above for: $y = -x^2 + x + 2$.

$y = -x^2 + x + 2$
when x = -3
$y = -(-3)^2 + -3 + 2$
$y = -9 - 3 + 2$
$y = -12 + 2$
y = -10

$y = -x^2 + x + 2$
when x = -2
$y = -(-2)^2 + -2 + 2$
$y = -4 - 2 + 2$
$y = -6 + 2$
y = -4

$y = -x^2 + x + 2$
when x = -1
$y = -(-1)^2 + -1 + 2$
$y = -1 - 1 + 2$
$y = -2 + 2$
y = 0

$y = -x^2 + x + 2$
when x = 0
$y = -(0)^2 + 0 + 2$
$y = 0 + 0 + 2$
y = 2

$y = -x^2 + x + 2$
when x = 1
$y = -(1)^2 + 1 + 2$
$y = -1 + 1 + 2$
y = 2

$y = -x^2 + x + 2$
when x = 2
$y = -(2)^2 + 2 + 2$
$y = -4 + 4$
y = 0

$y = -x^2 + x + 2$
when x = 3
$y = -(3)^2 + 3 + 2$
$y = -9 + 5$
y = -4

$y = -x^2 + x + 2$
when x = 4
$y = -(4)^2 + 4 + 2$
$y = -16 + 6$
y = -10

(8 marks)

(b) On the graph paper on the next page, draw the graph of $y = -x^2 + x + 2$ using the table above. Use an appropriate scale.

(6 marks)

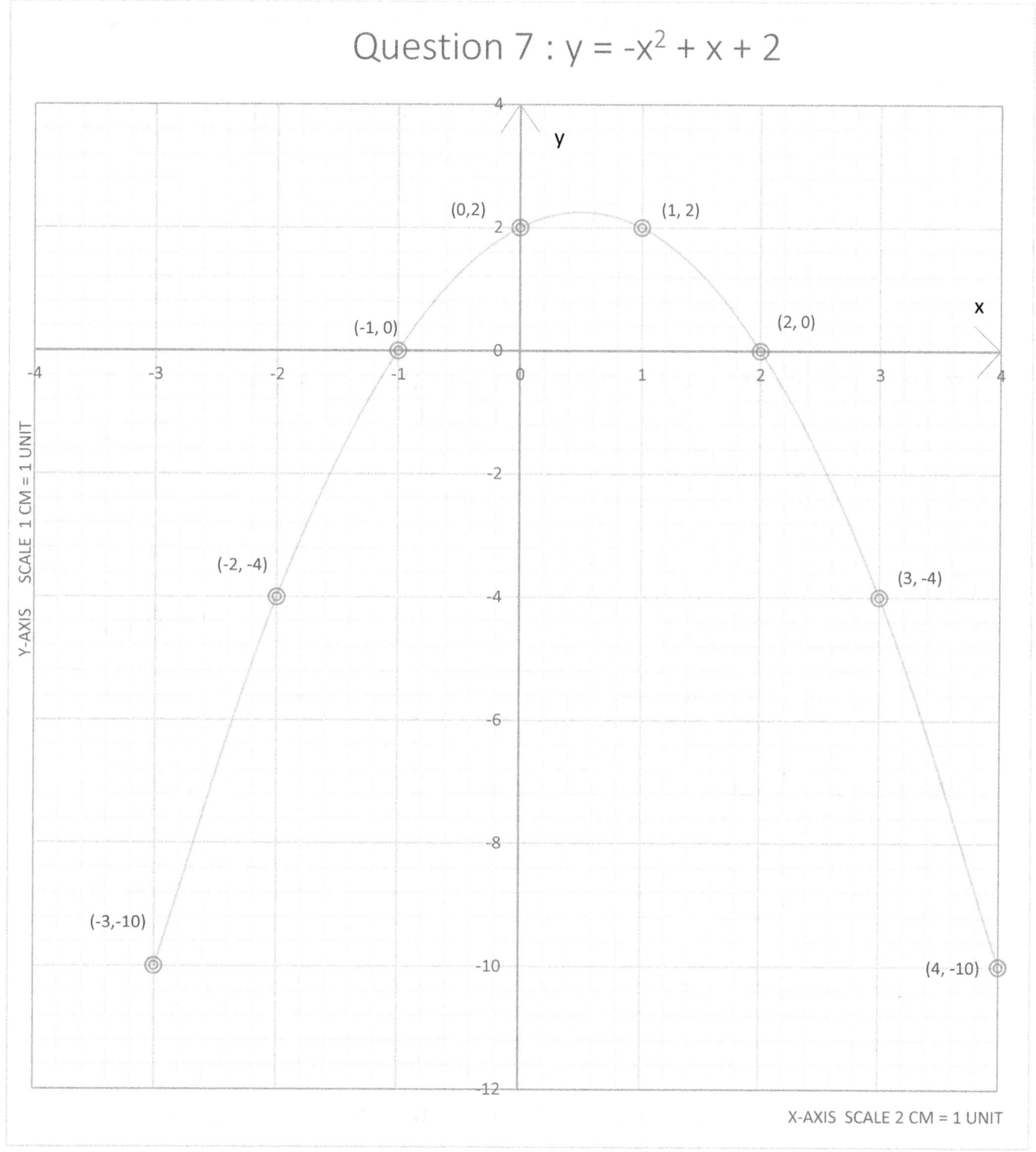

Complete the following statements using information from your graph.

(c) The values of x for which $-x^2 + x + 2 = 0$ occur are at ___-1___ and ___2___

(2 marks)

(d) The equation of the axis of symmetry of the graph $y = -x^2 + x + 2$ is ___$x = \frac{1}{2}$___

(2 marks)

(e) The maximum value of $-x^2 + x + 2$ is ___$2\frac{1}{4}$___

(1 mark)

(f) The name of the graph of a quadratic function is a ___parabola___

(1 mark)

END OF WORKSHEET

868

TUTORS

Preparation for

High School Mathematics

Sets

Solutions

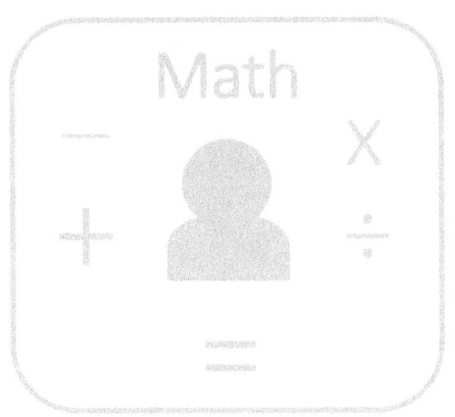

Instructions and Tips:

- ✓ You have 75 minutes to complete this worksheet
- ✓ This worksheet consists of 7 questions
- ✓ Write answers in the spaces provided
- ✓ All working must be clearly shown

Student Name: _____

Student ID: _____

Date: __/__/____

Total Score:

Highest Score:

Tutor's Comments:

Access more free worksheets at www.868tutors.com

Question 1

The Venn diagram below illustrates the number of students who drink juice (J) or carbonated beverages (C) during lunch time in a school of 500 students.

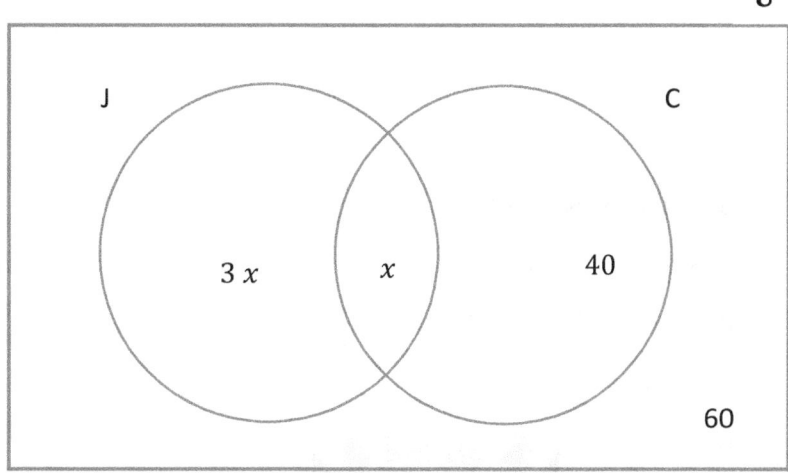

(a) How many students drink neither Juice nor Carbonated beverages during lunch time?

From Venn diagram

n (J ∪ C)' = 60

60 students drink neither Juice nor Carbonated beverages during lunch time

(1 mark)

(b) Write an expression, in terms of x, which represents the TOTAL number of students in the school.

$3x + x + 40 + 60$

(1 mark)

(c) Write an equation which may be used to determine the total number of students who drink both juice and carbonated beverages during lunch time.

We need an equation in terms of x

$$\boxed{3x + x + 40 + 60 = 500}$$

(1 mark)

(d) How many students drink juice?

Students that drink juice= Students that drink juice only + students that drink both juice and carbonated beverages

Number of students that drink juice = $3x + x = 4x$

Recall $3x + x + 40 + 60 = 500$ (c)

$4x = 500 - 100$, $4x = 400$, $x = 100$

Number of students that drink juice = 400

(1 mark)

(e) How many students drink carbonated beverages only?

From Venn diagram

Number of students that drink carbonated beverages only = 40

(1 mark)

Question 2

The Venn diagram below illustrates the number of people who make their own ice-cream (M) or purchase ice-cream (P) in a town of 1000 people.

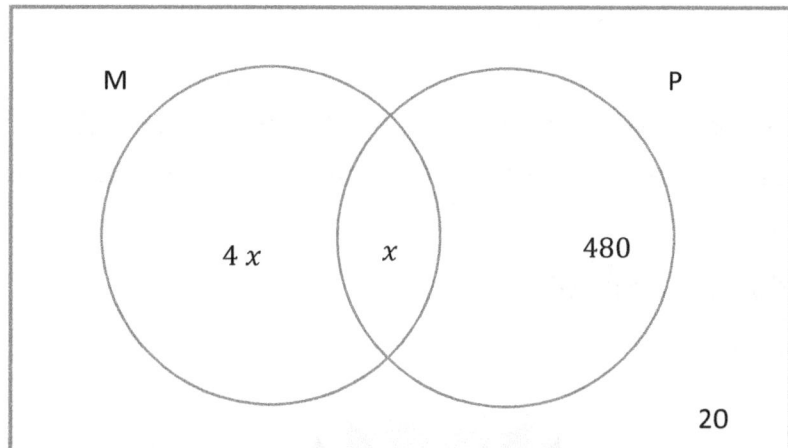

(a) How many people in the town neither make their own ice-cream nor purchase ice-cream?

From Venn diagram

n (M ∪ P)′ = 20

20 people in the town neither make their own ice-cream nor purchase ice-cream

(1 mark)

(b) Write an expression, in terms of x, which represents the TOTAL number of people in the town.

$\boxed{4x + x + 480 + 20}$

(1 mark)

(c) Write an equation which may be used to determine the total number of people in the town who make their own ice-cream and purchase ice-cream.

We need an equation in terms of x

$\boxed{4x + x + 480 + 20 = 1000}$

(1 mark)

(d) How many people in the town make their own ice-cream?

People who make their own ice-cream = People who make ice-cream only + People who make ice-cream and purchase ice-cream

People who make their own ice-cream = $4x + x = 5x$

Recall from (c) $4x + x + 480 + 20 = 1000$

$5x + 500 = 1000$, $x = 100$

$5x = 500$ $\boxed{\text{500 people make their own ice-cream}}$

(1 mark)

(e) How many people in the town purchase ice-cream only?

From the Venn Diagram

$\boxed{\text{Number of people in town who purchase ice cream only = 480}}$

(1 mark)

Question 3

In a town of 50 households,

30 households own pickup trucks,

3y students own compact cars only

y households own both compact cars and pickup trucks

2 households own neither compact cars nor pickup trucks

(a) Draw and clearly label a Venn diagram to illustrate the aforementioned information

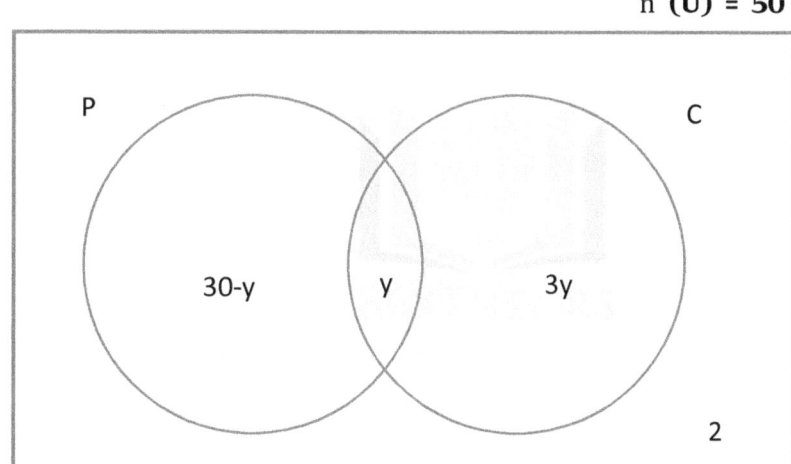

Let P represent Pickup trucks

Let C represent Compact Cars

(3 marks)

(b) Calculate the number of households that own compact cars

Owners of compact cars = owners of compact cars only + owners of compact cars and pickup trucks

Owners of compact cars = 3y + y = 4y

We need to write an equation in terms of y and solve for y

30-y + y + 3y + 2 = 50, 3y + 32 = 50, 3y = 18, y = 6

4y = 24, **24 households own compact cars**

(2 marks)

(c) Calculate the number of households that own pickup trucks only

From the Venn diagram

Number of households that own pickup trucks only = 30 - y

From (b) y = 6

30 - 6 = 24

| **24 households own pickup trucks only** |

(2 marks)

Question 4

In a class of 35 students,

10 students have visited the Erin mud volcanoes

3z students have visited the La Brea Pitch Lake only

z students have visited both the Erin mud volcanoes and the La Brea Pitch Lake

1 person has visited neither the Erin Mud Volcanoes nor the La Brea Pitch Lake

(a) Draw and clearly label a Venn diagram to illustrate the aforementioned information

n (U) = 35

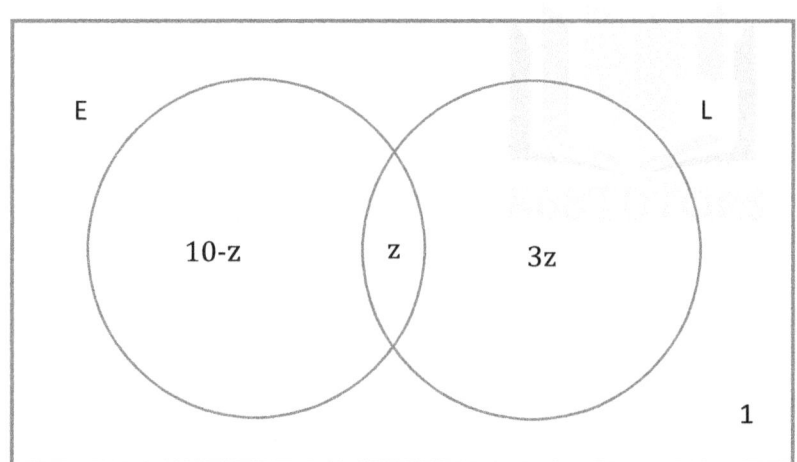

Let E represent visitors to the Erin Mud Volcanoes

Let L represent visitors to the La Brea Pitch Lake

(2 marks)

(b) Calculate the number of students that have visited the La Brea Pitch Lake

Visitors to the La Brea Pitch Lake = Visitors to the La Brea Pitch Lake only + Visitors to the La Brea Pitch Lake and the Erin Mud Volcanoes

Visitors to the La Brea Pitch Lake = 3z + z = 4z

We need to write an equation in terms of z (10-z +z+3z+1 =35)

3z + 11= 35

3z = 35-11   3z = 24   z = 8   (Therefore, 4z = 32)

32 students have visited the La Brea Pitch Lake

(2 marks)

(c) Calculate the number of students that have visited the Erin mud volcanoes only

From the Venn diagram

Number of students that have visited the Erin Mud Volcanoes only = 10 - z

Recall from (b) z = 8

Number of students that have visited the Erin Mud Volcanoes only = 10 - 8 = 2

2 students have visited the Erin Mud Volcanoes only

(2 marks)

Question 5

The universal set, U is defined as:

U = {10,11,12,13,14,15,16,17,18,19,20}

X and Y are subsets of U, such that:

X = {even numbers}

Y = {odd numbers}

(a) List the members of the set X

X = {10, 12, 14, 16, 18, 20}

(1 mark)

(b) List the members of the set Y

Y = {11, 13, 15, 17, 19,}

(1 mark)

Draw a Venn diagram to represent the sets X, Y and U

(c)

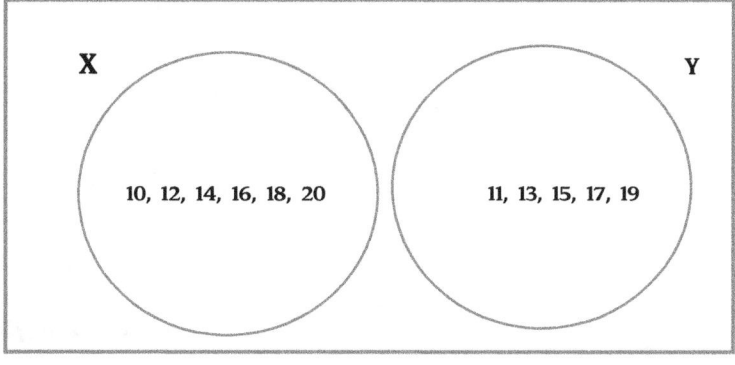

(3 marks)

Question 6

The Venn diagram below illustrates the number of people who drink coconut water (C) or spring water (S) in a village of 100 people.

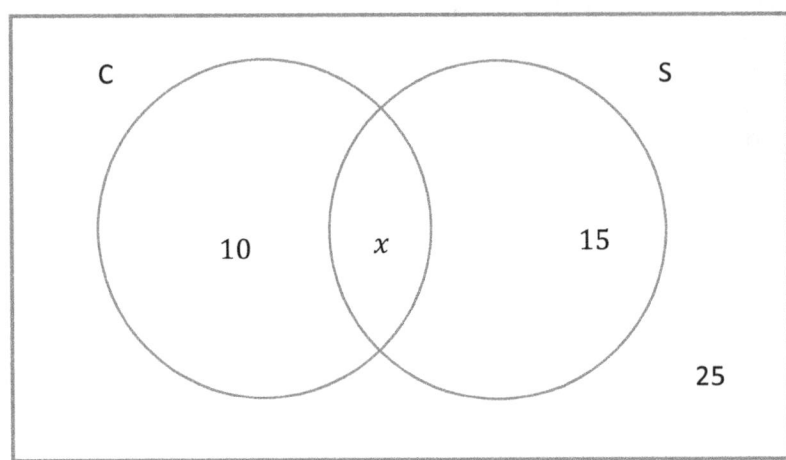

(a) How many people in the village drink neither coconut water nor spring water?

n (C ∪ S)′ = 25

25 people in the village drink neither coconut water nor spring water

(1 mark)

(b) Write an expression, in terms of x, which represents the total number of people in the village.

10 + *x* + 15 + 25

(1 mark)

(c) Write an equation which may be used to determine the total number of people in the village who drink coconut water and spring water.

10 + *x* + 15 + 25 = 100

(1 mark)

(d) How many people in the village drink coconut water?

People who drink coconut water = People who drink coconut water only + People who drink coconut water and spring water

People who drink coconut water = 10 + *x*

We need to write an equation in terms of x and solve for x

10 + *x* + 15 + 25 = 100

x + 50 = 100, *x* = 100 - 50 , *x* = 50

Therefore, 10 + *x*, 60 people drink coconut water

(1 mark)

(e) How many people in the village drink spring water only?

From the Venn diagram

15 people drink spring water only

(1 mark)

Question 7

The Venn diagram below illustrates the number of people who plant vegetables (K) and those who purchase vegetables (P) in a community of 120 people.

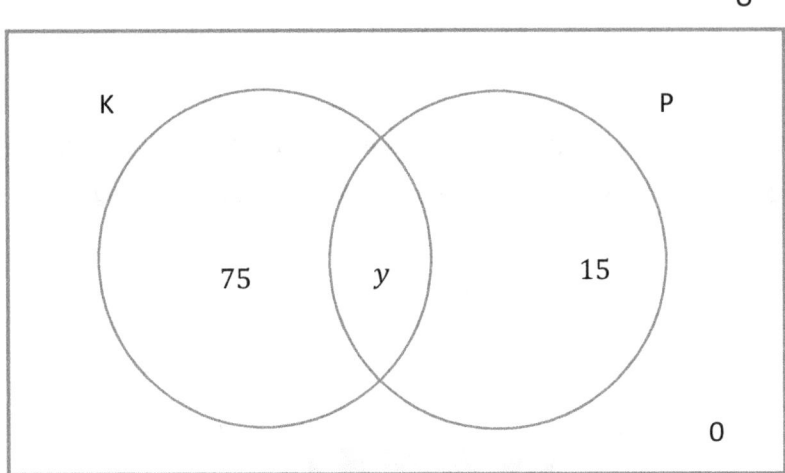

(a) How many people in the community neither plant their own vegetables nor purchase vegetables?

From the Venn diagram

$n(K \cup P)' = 0$

0 people in the community neither plant their own vegetables nor purchase vegetables

(1 mark)

(b) Write an expression, in terms of y, which represents the total number of people in the village.

$75 + y + 15 + 0$

$75 + y + 15$

(1 mark)

(c) Write an equation which may be used to determine the total number of people in the village who plant vegetables and purchase vegetables.

$\boxed{75 + y + 15 = 120}$

(1 mark)

(d) Write an expression in y for the number of people who plant their own vegetables

People who plant their own vegetables = People who plant their own vegetables only + People who plant vegetables and purchase vegetables

$\boxed{\text{People who plant their own vegetables} = 75 + y}$

(1 mark)

(e) How many people in the village purchase vegetables only?

From the Venn diagram

$\boxed{\text{15 people purchase vegetables only}}$

(1 mark)

END OF WORKSHEET

868 TUTORS

Preparation for
High School Mathematics
Simultaneous Equations
Solutions

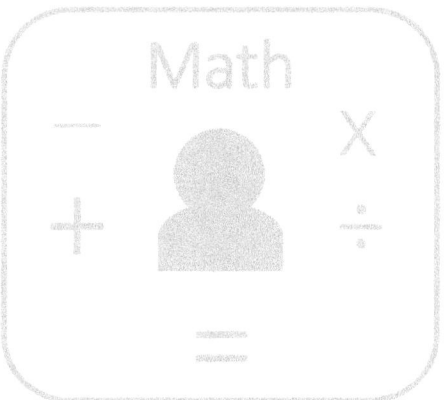

Instructions and Tips:

- ✓ You have 60 minutes to complete this worksheet
- ✓ This worksheet consists of 7 questions
- ✓ Write answers in the spaces provided
- ✓ All working must be clearly shown
- ✓ Label Graphs properly

Student Name: _____

Student ID: _____

Date: __/__/____

Total Score:

Highest Score:

Tutor's Comments:

Access more free worksheets at www.868tutors.com

Question 1

Solve the simultaneous equations:

2x + 5y = 16

x + 4y = -1

$2x + 5y = 16$ (i)

$x + 4y = -1$ (ii)

Rearrange equation (ii)

$x + 4y = -1$

$x = -1 - 4y$ (iii)

Substitute equation (iii) into equation (i)

$2(-1 - 4y) + 5y = 16$

$-2 - 8y + 5y = 16$

$-2 - 3y = 16$

$-3y = 16 + 2$

$-3y = 18$

$y = \dfrac{18}{-3}$

$y = -6$

Substitute y = -6 into equation (iii)

$x = -1 - 4(-6)$

$x = -1 + 24$

$x = 23$

x = ____23____

y = ____-6____

(4 marks)

Question 2

Solve the simultaneous equations:

x + y = 11

2x - 5y = -13

$x + y = 11$ (i)

$2x - 5y = -13$ (ii)

Rearrange equation (i)

$x + y = 11$

$y = 11 - x$ (iii)

Substitute equation (iii) into equation (ii)

$2x - 5(11 - x) = -13$

$2x - 55 + 5x = -13$

$2x + 5x = -13 + 55$

$7x = 42$

$x = \frac{42}{7} = 6$

Substitute $x = 6$ into equation (iii)

$y = 11 - 6$

$y = 5$

x = _____6_____

y = _____5_____

(4 marks)

Question 3

Solve the simultaneous equations:

2x + y = 12

x - y = - 6

$2x + y = 12$ (i)

$x - y = -6$ (ii)

Rearrange equation (i)

$2x + y = 12$

$y = 12 - 2x$ (iii)

Substitute equation (iii) into equation (ii)

$x - (12 - 2x) = -6$

$x - 12 + 2x = -6$

$x + 2x - 12 = -6$

$3x - 12 = -6$

$3x = -6 + 12$

$3x = 6$

$x = \frac{6}{3}$ $x = 2$

Substitute $x = 2$ into equation (iii)

$y = 12 - 2(2)$ (iii)

$y = 12 - 4$

$y = 8$

x = _____2_____

y = _____8_____   (4 marks)

Question 4

Solve the simultaneous equations:

$3x^2 + y^2 = 6$

$3x - y = -3$

$3x^2 + y^2 = 6$ (i)

$3x - y = -3$ (ii)

Rearrange equation (ii)

$3x - y = -3$

$-y = -3 - 3x$ (divide throughout by -1)

$y = 3 + 3x$ (iii)

Substitute equation (iii) into equation (i)

$3x^2 + (3 + 3x)^2 = 6$

$3x^2 + (3 + 3x)(3 + 3x) = 6$

$3x^2 + 9 + 9x + 9x + 9x^2 = 6$

$12x^2 + 18x + 3 = 0$

÷3

$4x^2 + 6x + 1 = 0$

$a = 4 \quad b = 6 \quad c = 1$

Applying the quadratic formula

$x = \dfrac{-b \pm \sqrt{b^2 - 4ac}}{2a}$

$x = \dfrac{-6 + \sqrt{6^2 - 4(4)(1)}}{2(4)}$

$x = \dfrac{-6 + \sqrt{36 - 16}}{8}$

$x = \dfrac{-6 + \sqrt{20}}{8}$

x = -0.190983005

Substitute x value into equation (iii)

y = 3 + 3(-0.190983005) = 2.427050983

$x = \dfrac{-6 - \sqrt{36 - 16}}{2(4)}$

$x = \dfrac{-6 - \sqrt{36 - 16}}{8}$

$x = \dfrac{-6 - \sqrt{20}}{8}$

x = -1.309016994

Substitute x value into equation (iii)

y = 3 + 3(-1.309016994) = -0.927050983

x = _____-0.19 or -1.31_____ (to 2 decimal places)

y = _____2.43   or   -0.93_____ (to 2 decimal places)

(6 marks)

Question 5

Solve the simultaneous equations:

2x - 3y = -14

x + 2y = 7

$2x - 3y = -14$ (i)

$x + 2y = 7$ (ii)

Rearrange equation (ii)

$x = 7 - 2y$ (iii)

Substitute equation (iii) into equation (i)

$2(7 - 2y) - 3y = -14$

$14 - 4y - 3y = -14$

$14 - 7y = -14$

$-7y = -14 - 14$

$-7y = -28$

$y = \frac{-28}{-7}$

$y = 4$

Substitute y = 4 into equation (iii)

$x = 7 - 2(4)$

$x = 7 - 8$

$x = -1$

x = _____-1_____

y = _____4_____

(4 marks)

Question 6

Solve the simultaneous equations:

x + y = 4

3x + y = 5

$x + y = 4$ (i)

$3x + y = 5$ (ii)

Rearrange equation (i)

$y = 4 - x$ (iii)

Substitute equation (iii) into equation (ii)

$3x + 4 - x = 5$

$2x + 4 = 5$

$2x = 1$

$x = \frac{1}{2}$

Substitute $x = \frac{1}{2}$ into equation (iii)

$y = 4 - \frac{1}{2}$

$y = 3\frac{1}{2}$

x = $\frac{1}{2}$

y = $3\frac{1}{2}$

(4 marks)

Question 7

Consider the simultaneous equations:

y = 2x + 2

y = -2x + 3

(a) Complete the table below for the equation: y = 2x + 2

x	-3	-2	-1	0	1	2
y	-4	-2	0	2	4	6

when x = -3, y = 2 (-3) + 2 = -4   when x = -2, y = 2 (-2) + 2 = -2   when x = -1, y = 2 (-1) + 2 = 0

when x = 0, y = 2 (0) + 2 = 2   when x = 1, y = 2 (1) + 2 = 4   when x = 2, y = 2 (2) + 2 = 6

(b) Complete the table below for the equation: y = -2x + 3

x	-3	-2	-1	0	1	2
y	9	7	5	3	1	-1

when x = -3, y = -2 (-3) + 3 = 9   when x = -2, y = -2 (-2) + 3 = 7   when x = -1, y = -2 (-1) + 3 = 5

when x = 0, y = -2 (0) + 3 = 3   when x = 1, y = -2 (1) + 3 = 1   when x = 2, y = -2 (2) + 3 = -1

(c) Draw the graphs of both equations on the same graph paper

(d) Use your graphs to solve the simultaneous equation

x = $\frac{1}{4}$

y = $2\frac{1}{2}$

(10 marks)

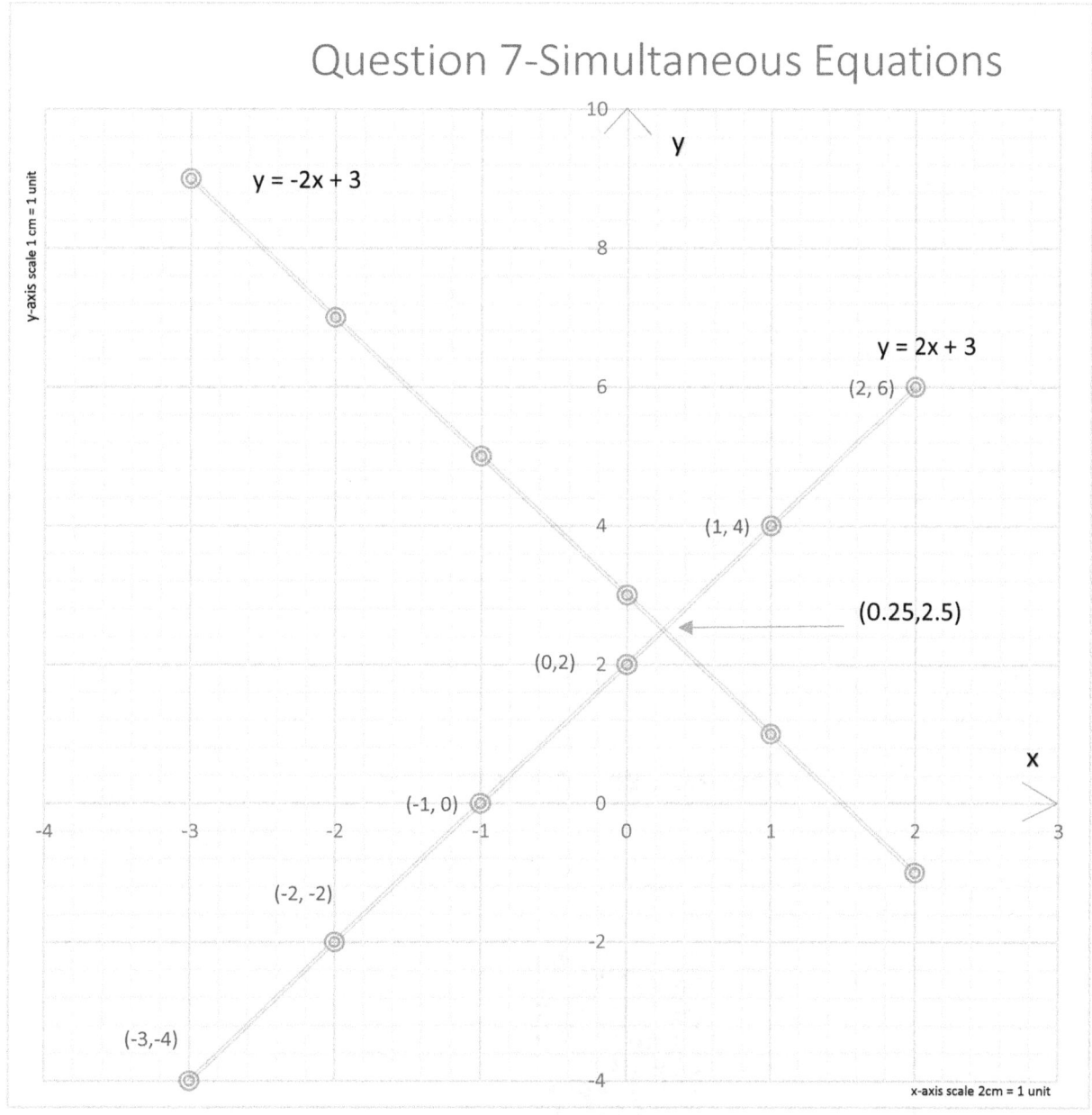

Question 7-Simultaneous Equations

The point of intersection of the two graphs is the solution to the simultaneous equations.

END OF WORKSHEET

868 TUTORS

Preparation for

High School Mathematics

Statistics
(Histograms and Line Graphs)

Solutions

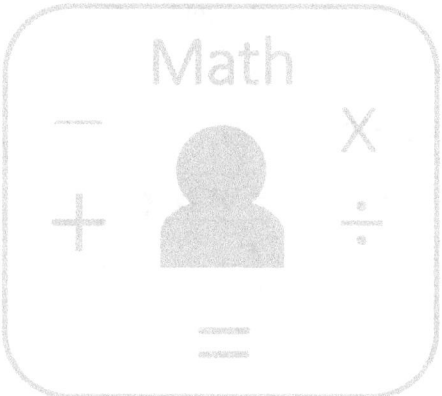

Instructions and Tips:

- ✓ You have 60 minutes to complete this worksheet
- ✓ This worksheet consists of 5 questions
- ✓ Write answers in the spaces provided
- ✓ All working must be clearly shown

Student Name: _____

Student ID: _____

Date: __/__/____

Total Score:

Highest Score:

Tutor's Comments:

Access more free worksheets at www.868tutors.com

Question 1

Twenty-five bags of Cocoa powder are measured on a scale. The mass of each bag is recorded to the nearest kilogram as shown in the table below:

2	35	6	22	6
4	42	5	26	14
13	7	47	19	32
50	20	43	14	34
49	9	41	13	38

(a) Complete the frequency table below for the given data.

Mass (kg)	Tally	Number of bags
1 - 10	⦀⦀⦀ ⦀⦀	7
11 - 20	⦀⦀⦀ ⦀	6
21 - 30	⦀⦀	2
31 - 40	⦀⦀⦀⦀	4
41 - 50	⦀⦀⦀ ⦀	6

(b) State the lower class boundary for the class interval 11 - 20.

          10.5

(c) State the class width for the class interval 11 – 20.

          $20.5 - 10.5 =$ **10**

(d) State the class midpoint for the class interval 11 – 20.

        $\frac{11+20}{2}$ **15.5**   (3 marks)

(e) On the graph paper on the next page, draw a histogram to represent the data contained in the frequency table above. Use appropriate scales (kilograms on the x–axis and bags on the y–axis).

                                                    (6 marks)

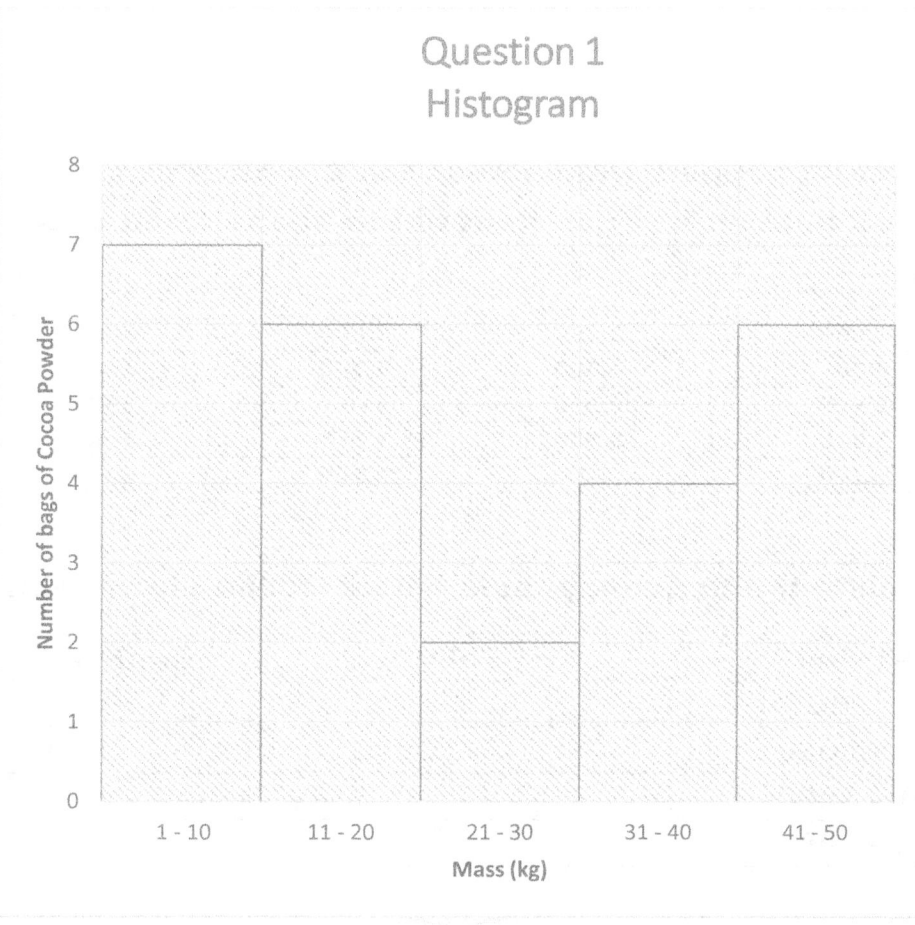

Appropriate scale : x-axis, 2 cm = 10 kg

Appropriate scale : y-axis, 1 cm = 1 bag

Question 2

Twenty different peppers from across the Caribbean were classified in terms of their heat units. Their heat to the nearest unit was recorded as shown:

100	3,800	1,800	300
900	1,300	1,700	400
2,500	4,000	110	3,900
3,500	3,600	1,600	3,500
1,200	3,900	3,100	3,050

(a) Complete the frequency table below for the given data.

Heat Unit	Tally	Number of peppers
1 - 1000	|||||	5
1001 - 2000	|||||	5
2001 - 3000			1
3001 - 4000	||||| ||||	9

(b) State the lower class boundary for the class interval 1001 - 2000.

   1000.5

(c) State the class width for the class interval 1001 - 2000.

  $2000.5 - 1000.5 = \boxed{1000}$

(d) State the class midpoint for the class interval 1001 – 2000.

  $\frac{1001 + 2000}{2} = \boxed{1500.5}$

                     (3 marks)

(e) On the graph paper on the next page, draw a histogram to represent the data contained in the frequency table above. Use appropriate scales (heat units on the x–axis and peppers on the y–axis).

                      (6 marks)

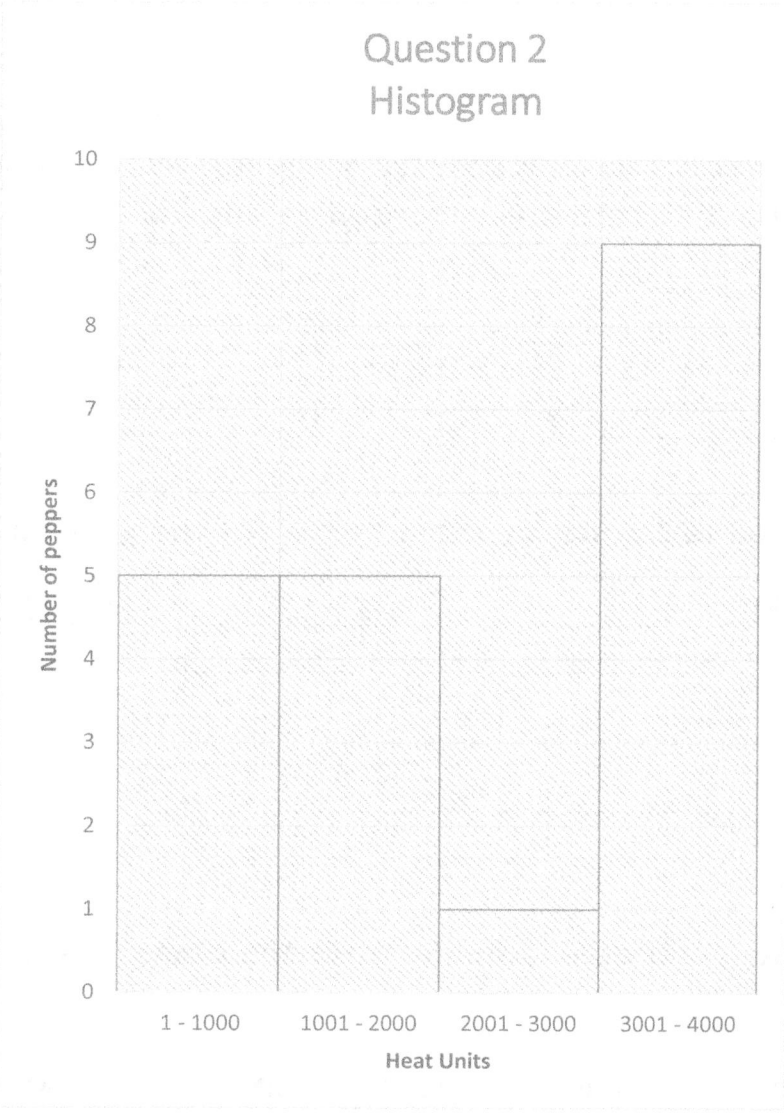

Appropriate scale : x-axis, 2 cm = 1000 Heat Units

Appropriate scale : y-axis, 1 cm = 1 pepper

Question 3

Twenty bags of coconut milk powder are weighed. The mass of each bag to the nearest kilogram is recorded as shown:

8	9	16	28
19	48	18	9
15	32	14	50
13	34	39	15
21	7	17	1

(a) Complete the frequency table below for the given data.

Mass (kg)	Tally	Number of bags
1 - 10	|||||	5
11 - 20	||||| |||	8
21 - 30	||	2
31 - 40	|||	3
41 - 50	||	2

(b) State the lower class boundary for the class interval 31 - 40.

30.5

(c) State the class width for the class interval 31 – 40.

$40.5 - 30.5 = \boxed{10}$

(d) State the class midpoint for the class interval 31 – 40.

$\frac{31+40}{2} = \boxed{35.5}$

(3 marks)

(e) On the graph paper on the next page, draw a histogram to represent the data contained in the frequency table above. Use appropriate scales (kilograms on the x–axis and bags on the y–axis.

(6 marks)

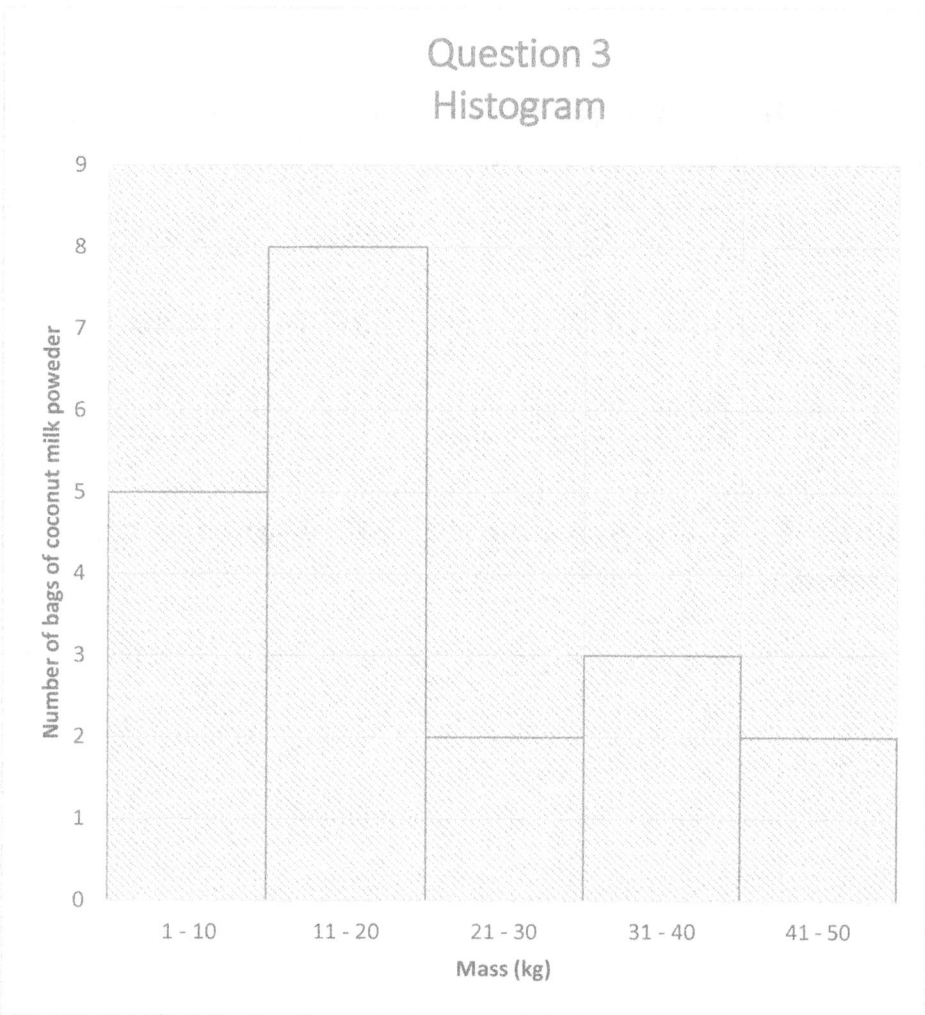

Appropriate scale : x-axis, 2 cm = 10 kg

Appropriate scale : y-axis, 1 cm = 1 bag

Question 4

The table below shows the quantity of hot peppers produced by a farm, in tonnes, from 2012 to 2016.

Year	2012	2013	2014	2015	2016
Hot pepper production (Tonnes)	8	5	9	10	12

(a) Complete the line graph below to represent the given information.

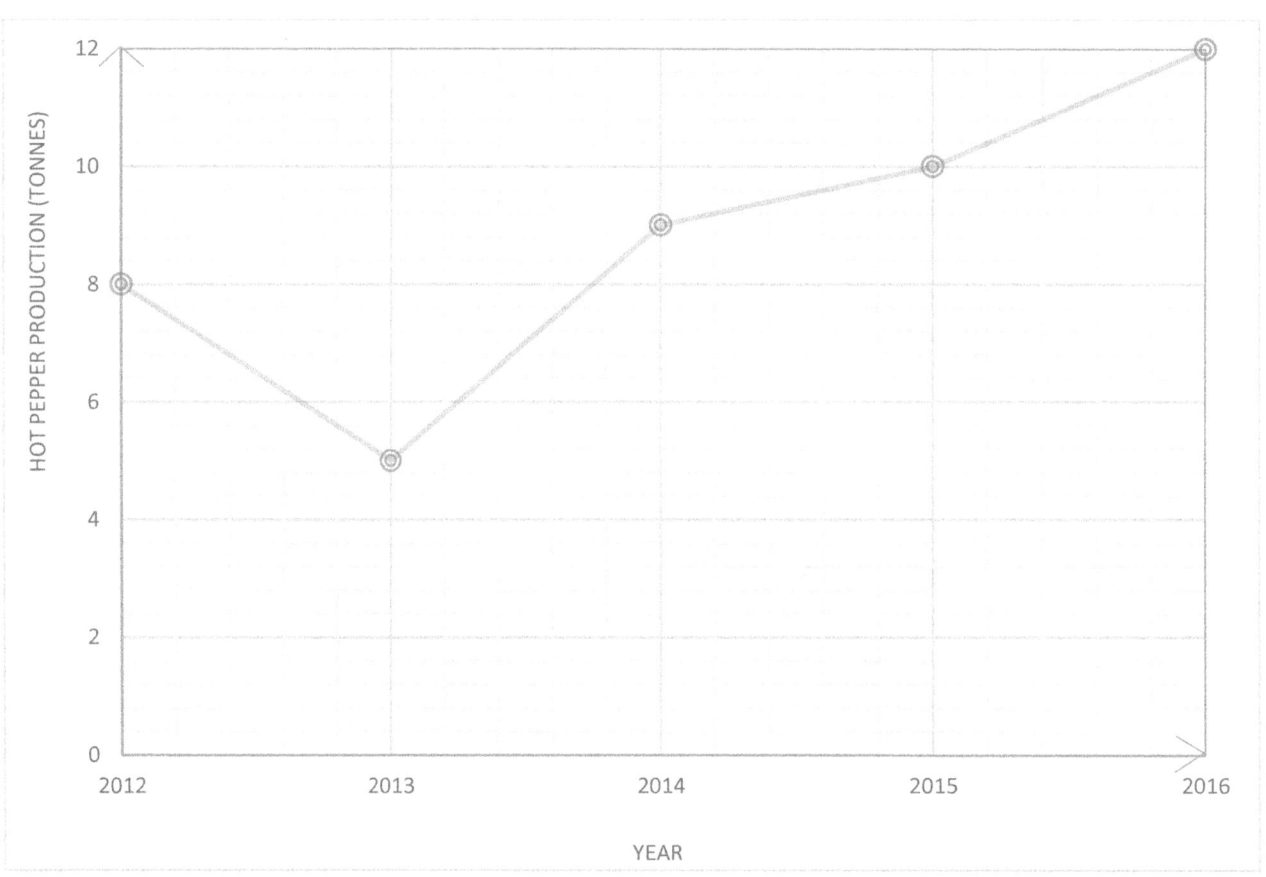

(2 marks)

(b) Between which two consecutive years was there the greatest increase in hot peppers produced?

_____2013_____ and _____2014_____

(the section of the line graph with the largest gradient indicates the greatest increase)

(c) What was the total production of hot peppers in the five year period from 2012 to 2016?

8 + 5 + 9 + 10 + 12 = $\boxed{44 \text{ tonnes}}$

(3 marks)

(d) The mean yearly hot pepper production from 2011 to 2016 was 10 tonnes. How many hot peppers were produced in 2011?

$$\frac{x + 8+5+9+10+12}{6} = \frac{10}{1}$$

$\frac{x + 44}{6} = \frac{10}{1}$ (Cross-Multiplying)

x + 44 = 60

x = 60 - 44

x = 16 tonnes

$\boxed{\text{16 tonnes of hot pepper were produced in 2011.}}$

(3 marks)

Question 5

The line graph below shows the daily production of a private oil producer, in hundreds of barrels over a five day period.

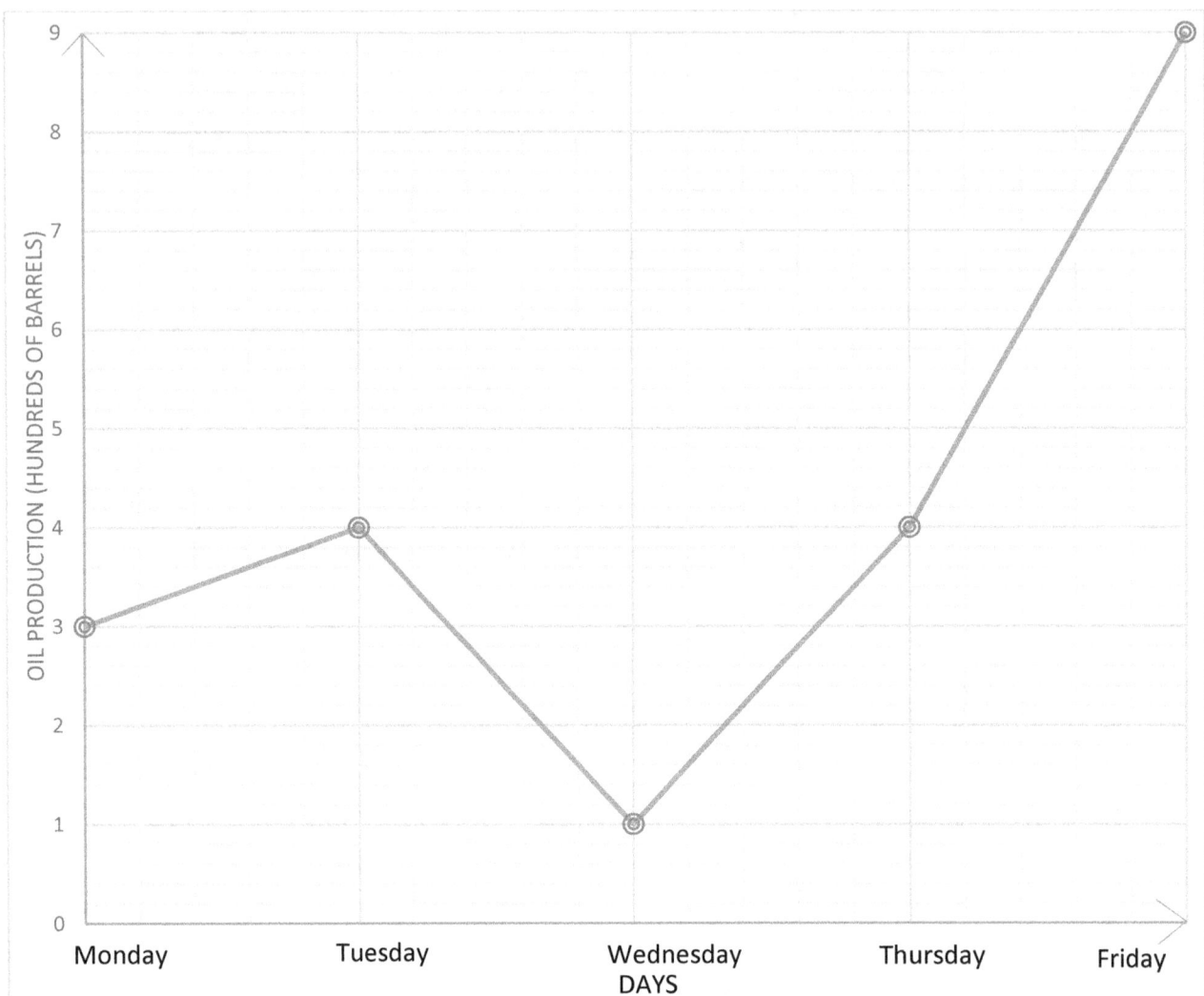

Complete the table to show the daily oil production.

Day	Monday	Tuesday	Wednesday	Thursday	Friday
Oil production (Hundreds of barrels)	3	4	1	4	9

(3 marks)

(a) Between which two consecutive days was there the greatest increase in oil production?

   Thursday   **and**   Friday

(b) What is the total oil production for the five day period, Monday to Friday?

   300 + 400 + 100 + 400 + 900 = 2100 barrels

(c) Calculate the mean daily oil production for the five day period from Monday to Friday.

$$\frac{300 + 400 + 100 + 400 + 900}{5} = 420 \text{ barrels}$$

$$\frac{\text{Total Number of Barrels}}{\text{Total Number of Days}} = \text{Mean daily oil production}$$

(4 marks)

END OF WORKSHEET

868 TUTORS

868

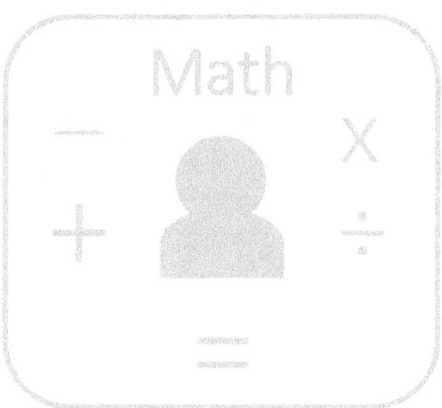

Preparation for

High School Mathematics

Statistics II
(Cumulative Frequency Curves and Pie Charts)
Solutions

Instructions and Tips:

- ✓ You have 90 minutes to complete this worksheet
- ✓ This worksheet consists of 6 questions
- ✓ Write answers in the spaces provided
- ✓ All working must be clearly shown

Student Name: _____

Student ID: _____

Date: __ / __ / ____

Total Score:

Highest Score:

Tutor's Comments:

Access more free worksheets at www.868tutors.com

Question 1

The table below illustrates the money spent on market goods to the nearest dollar, for persons visiting a vegetable stall.

Money spent in Trinidad and Tobago Dollars (TTD)	Number of persons	Cumulative Frequency
20-29	7	7
30-39	9	7 + 9 = 16
40-49	10	16 + 10 = 26
50-59	20	26 + 20 = 46
60-69	15	46 + 15 = 61
70-79	8	61 + 8 = 69
80-89	5	69 + 5 = 74

(a) Complete the table above

(2 marks)

(b) Using appropriate scales (TTD on the x-axis and persons on the y-axis), draw the cumulative frequency curve for this data on the provided graph paper.

x	y
19.5	0
29.5	7
39.5	16
49.5	26
59.5	46
69.5	61
79.5	69
89.5	74

Table: Points to be plotted

(4 marks)

(b)

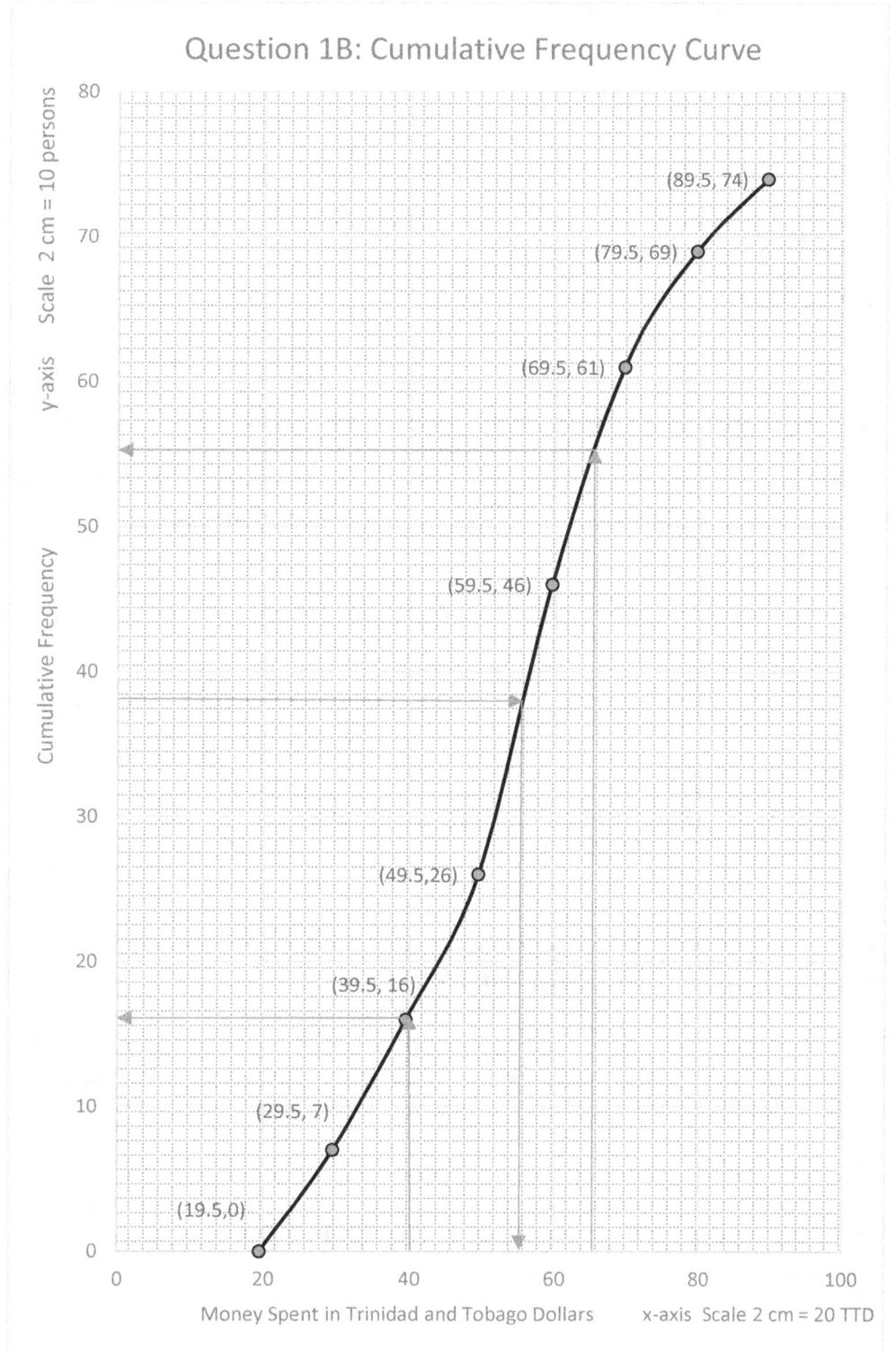

(c) Use the Cumulative Frequency Curve that you have drawn to determine (estimate with the use of lines).

(i) the median cost

Final cumulative frequency = 74 $\frac{74}{2} = 37$

(Draw a horizontal line from 37 on the y-axis until you meet the curve. Then draw a vertical line downwards. This value is the median.)

Median cost = $56 TTD

(2 marks)

(ii) the number of people who spent at least $65 TTD on goods

(Draw a vertical line on the x-axis from $65 TTD upwards until you meet the curve. Then draw a horizontal line. This value is the number of people who spent less than $65 TTD.)

Number of people who spent less than $65 TTD on goods = 55

Number of people who spent at least $65 TTD on goods = Final cumulative frequency - people who spent less than $65 TTD on goods

Number of people who spent at least $65 TTD on goods = 74-55

Number of people who spent at least $65 TTD on goods = 19

(2 marks)

(iii) the probability that a person, picked at random from the group, spent less than $40 TTD

(Draw a vertical line from $40 TTD upwards until you meet the curve. Then draw a horizontal line. This value is the number of people who spent less than $40 TTD.)

Number of people who spent less than $40 TTD on goods = 17

Probability that someone spent less than $40 TTD on goods = $\frac{17}{74}$

(2 marks)

Question 2

The table below illustrates the waiting time to the nearest minute, at a popular fast-food restaurant in Rancho Quemado, Trinidad.

Waiting Time (minutes)	Number of persons	Cumulative Frequency	Points to be plotted (x,y)
1-5	20	20	(5.5,20)
6-10	25	20 + 25 = 45	(10.5,45)
11-15	30	45+ 30 = 75	(15.5,75)
16-20	35	75+ 35 = 110	(20.5,110)
21-25	15	110+ 15 = 125	(25.5,125)
26-30	5	125 + 5 = 130	(30.5,130)
31-35	1	130+ 1 = 131	(35.5,131)

(a) Complete the table shown above.

(2 marks)

(b) Using appropriate scales (minutes on the x-axis and persons on the y-axis, draw the cumulative frequency curve for this data on the provided graph paper.

x	y
0.5	0
5.5	20
10.5	45
15.5	75
20.5	110
25.5	125
30.5	130
35.5	131

(4 marks)

(b)

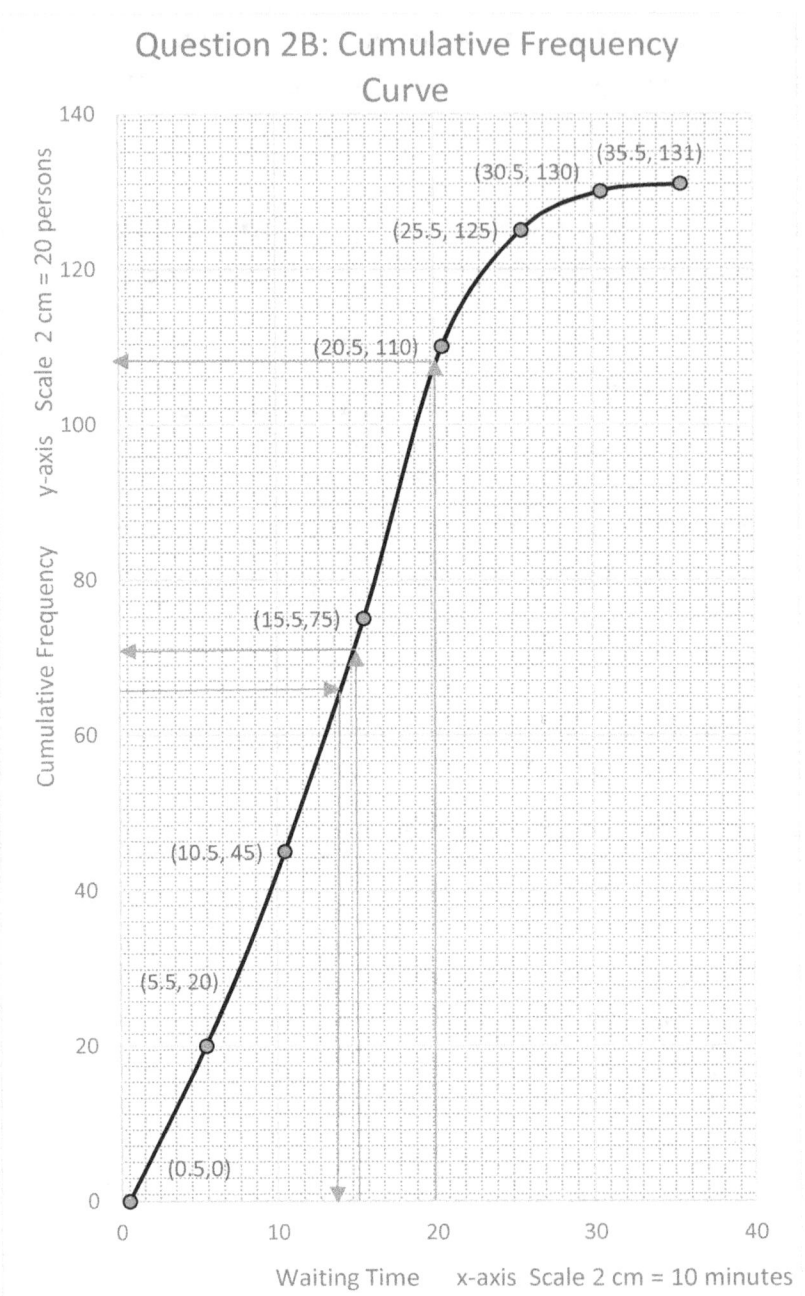

(c) Use the Cumulative Frequency Curve that you have drawn to determine (estimate with the use of lines).

(i) the median waiting time

Final cumulative frequency = 131 $\frac{131}{2}$ = 65.5

Draw a horizontal line from 65.5 on the y-axis until you meet the curve. Then draw a vertical line downwards. This value is the median.

Median waiting time = 14 minutes

(2 marks)

(ii) the number of people who waited at least 15 minutes

Draw a vertical line from 15 minutes upwards until you meet the curve. Then draw a horizontal line. This value is the number of people who spent less than 15 minutes.

Number of people who waited less than 15 minutes = 72

Number of people who waited at least 15 minutes = Final cumulative frequency-number of people who waited less than 15 minutes

Number of people who waited at least 15 minutes = 131-72 = 59

Number of people who waited at least 15 minutes = 59

(2 marks)

(iii) the number of people who waited at least 20 minutes

number of people who waited at least 20 minutes = Final Cumulative frequency- number who waited less than 20 minutes

number of people who waited at least 20 minutes = 131-106 = 25

(2 marks)

(iv) What is the name given to the shape of the curve that you have drawn?

The name given to the shape of the curve is: **Ogive**

(1 mark)

Question 3

The table below illustrates the number of hours (to the nearest hour) spent on academic activities in a Carnival long weekend for a group of secondary school students.

Number of hours (academic activities)	Number of persons	Cumulative Frequency	Points to be plotted (x,y)
1-3	1	1	(3.5,1)
4-6	3	1 + 3 = 4	(6.5,4)
7-9	5	4 + 5 = 9	(9.5,9)
10-12	20	9 + 20 = 29	(12.5, 29)
13-15	15	29 +15 = 44	(15.5.44)
16-18	5	44 + 5 = 49	(18.5.49)
19-21	2	49 + 2 = 51	(21.5,51)

(a) Complete the table shown above.

(2 marks)

(b) Use appropriate scales, (hours on the x-axis and persons on the y-axis), draw the cumulative frequency curve for this data on the provided graph paper.

x	y
0.5	0
3.5	1
6.5	4
9.5	9
12.5	29
15.5	44
18.5	49
21.5	51

(4 marks)

(b)

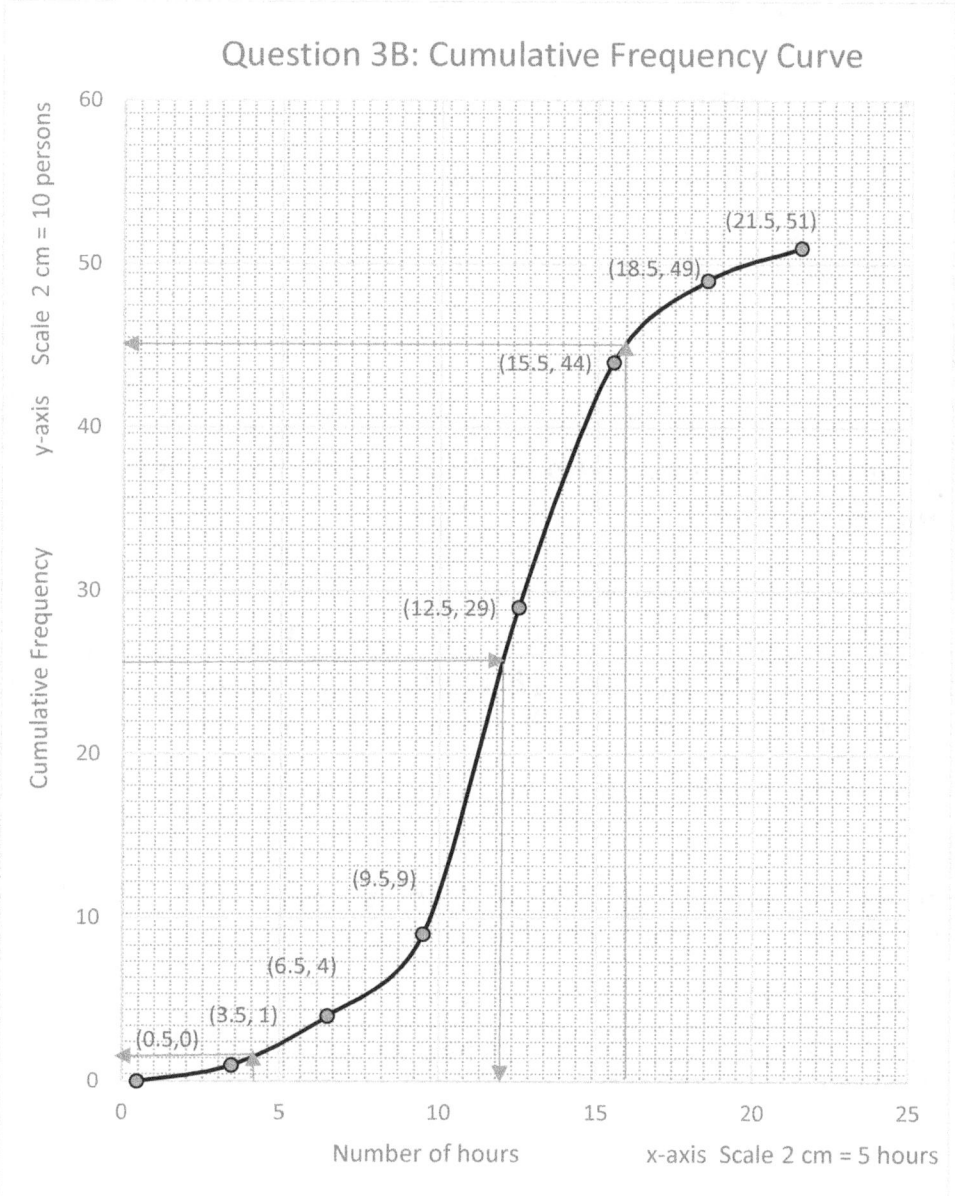

(c) Use the Cumulative Frequency Curve that you have drawn to determine (estimate with the use of lines).

(i) the median number of hours

Final cumulative frequency = 51 $\quad \frac{51}{2} = 25.5$

Draw a horizontal line from 25.5 on the y-axis until you meet the curve. Then draw a vertical line downwards. This value is the median

Median = 12 hours

(2 marks)

(ii) the number of people who dedicated at least 16 hours to academic activities

Draw a vertical line on the x-axis from 16 hours upwards until you meet the curve. Then draw a horizontal line. This value is the number of people who dedicated less than 16 hours on academic activities

Number of people who dedicated less than 16 hours = 45

Number of people who dedicated at least 16 hours = Final cumulative frequency- number of people who dedicated less than 16 hours

Number of people who dedicated at least 15 hours = 51-45 = 6

Number of people who dedicated at least 16 hours = 6

(2 marks)

(iii) the number of people who dedicated less than 4 hours to academic activities

Draw a vertical line on the x-axis from 4 hours upwards until you meet the curve. Then draw a horizontal line. This value is the number of people who spent less than 4 hours on academic activities

number of people who dedicated less than 4 hours to academic activities = 1

(2 marks)

Question 4

The table below illustrates the weekend plans of families in a community in southern Trinidad.

Weekend Plan	Number of families	Sector angle
Visit Los Iros Beach	10	48°
Visit Clifton Hill Beach	15	72°
Travel to Tobago	25	120°
Observe Carnival celebrations in Cedros	20	96°
Spend time at home	5	24°

(a) Calculate the corresponding sector angles required to draw a pie chart and complete the table.

Total number of families = 10 + 15 + 25 + 20 + 5 = 75

Sector angle (Visit Los Iros Beach) = $\frac{10}{75} \times 360° = 48°$

Sector angle (Visit Clifton Hill Beach) = $\frac{15}{75} \times 360° = 72°$

Sector angle (Travel to Tobago) = $\frac{25}{75} \times 360° = 120°$

Sector angle (Observe Carnival celebrations in Cedros) = $\frac{20}{75} \times 360° = 96°$

Sector angle (Spend time at home) = $\frac{5}{75} \times 360° = 24°$

(5 marks)

(b) Draw and label a pie chart to represent the information in the aforementioned table.

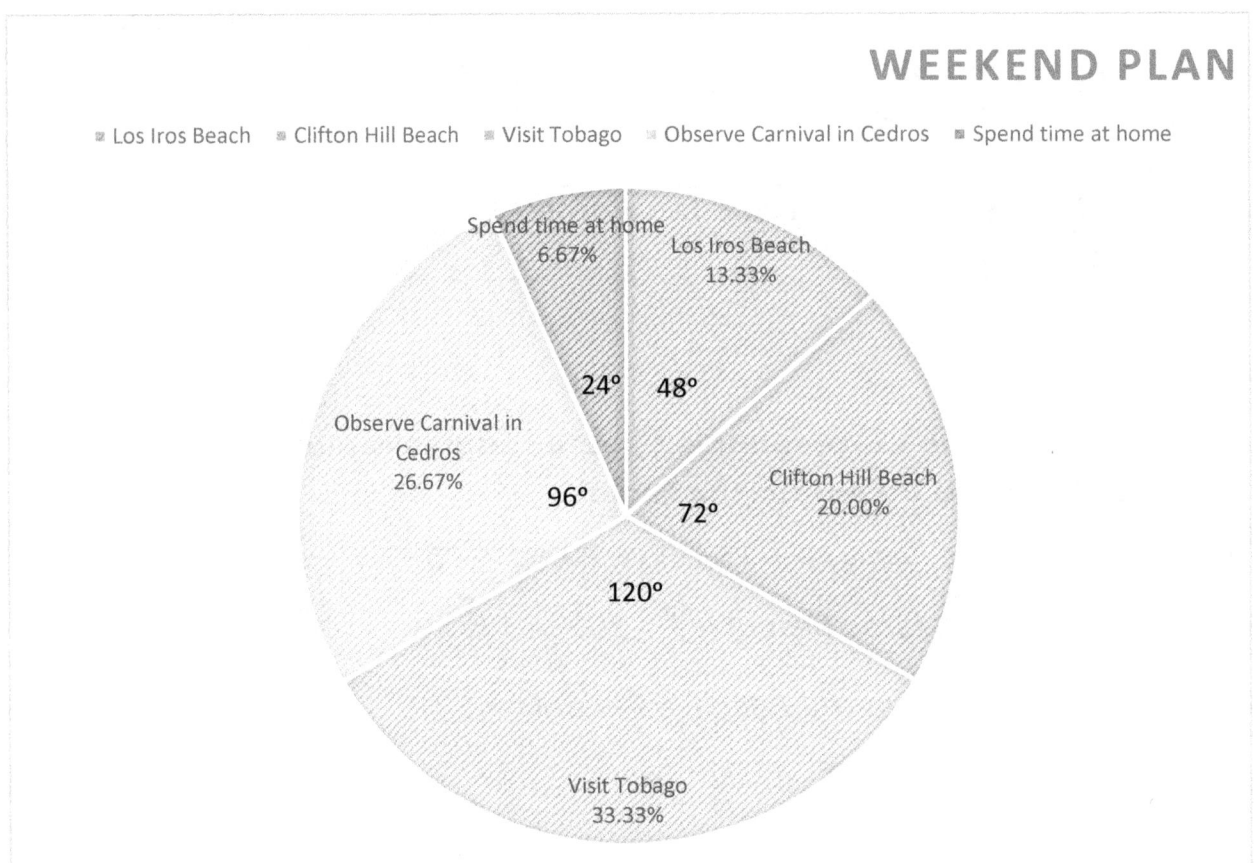

(5 marks)

Question 5

The information in the table below represents the favourite fast foods of a group of Form 6 students.

Food	Number of students	Sector angle
Doubles	30	135°
Fried Chicken	20	90°
Jerk Chicken	12	54°
BBQ Chicken	10	45°
Pizza	8	36°

(a) Calculate the sector angles required to draw a pie chart and complete the table.

Total number of students = 30 + 20 + 12 + 10 + 8 = 80

Sector angle (Doubles) = $\frac{30}{80} \times 360° = 135°$

Sector angle (Fried Chicken) = $\frac{20}{80} \times 360° = 90°$

Sector angle (Jerk Chicken) = $\frac{12}{80} \times 360° = 54°$

Sector angle (BBQ Chicken) = $\frac{10}{80} \times 360° = 45°$

Sector angle (Pizza) = $\frac{8}{80} \times 360° = 36°$

(5 marks)

(b) Draw and label a pie chart to represent the information in the aforementioned table.

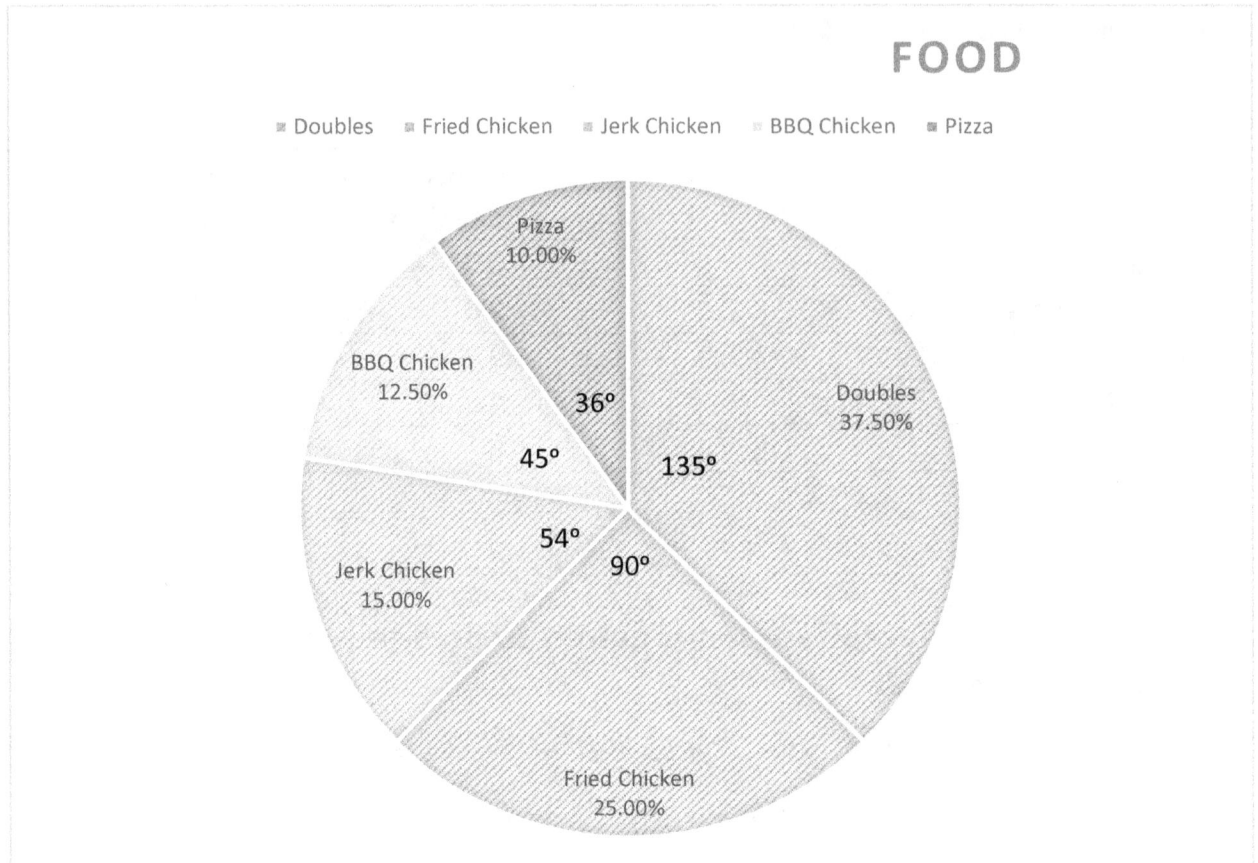

(5 marks)

Question 6

A group of 100 students were surveyed to determine their favourite fruits. The information is presented in the Pie Chart shown below:

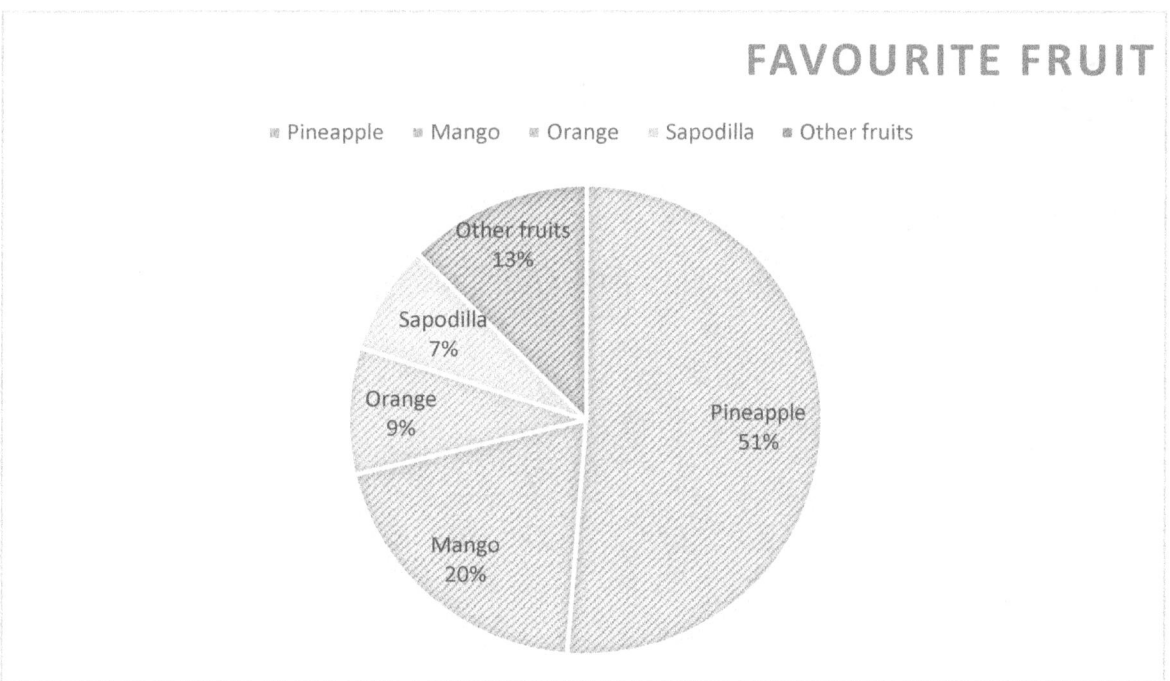

(a) Interpret the data and calculate the sector angle for the section of the pie chart that represents students who prefer mangoes.

20% prefer mangoes

$\frac{20}{100} \times 360° = 72°$

Sector angle for the section of the pie chart that represents students who prefer mangoes = 72°

(2 marks)

(b) Interpret the pie chart and complete the table (Show all working).

Favourite Fruit	Number of students
Pineapple	51
Mango	20
Orange	9
Sapodilla	7

Total number of students = 100

We observe from the given information (pie chart) and table that Pineapple = 51% = 51

Mango

$20\% = \frac{20}{100} \times 100 = 20$

Orange

$9\% = \frac{9}{100} \times 100 = 9$

Sapodilla

$7\% = \frac{7}{100} \times 100 = 7$

(3 marks)

END OF WORKSHEET

868

868TUTORS

TUTORS

Preparation for

High School Mathematics

Straight Line Graphs

Solutions

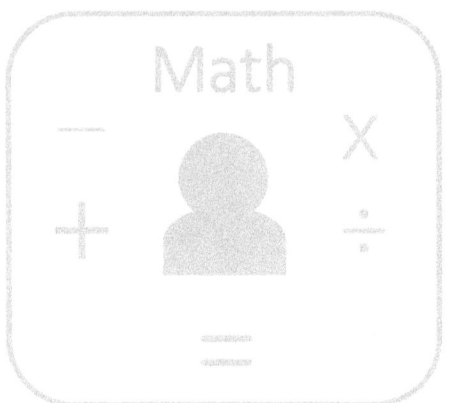

Instructions and Tips:

- ✓ You have 90 minutes to complete this worksheet
- ✓ This worksheet consists of 8 questions
- ✓ Write answers in the spaces provided
- ✓ All working must be clearly shown
- ✓ Label Graphs properly

Student Name: _____

Student ID: _____

Date: __/__/____

Total Score:

Highest Score:

Tutor's Comments:

Access more free worksheets at www.868tutors.com

Question 1

Consider the straight line equation: y = x + 1.

x	-3	-2	-1	0	1	2
y	-2	-1	0	1	2	3

(a) Complete the table above for: y = x + 1.

y = x + 1
when x = -3
y = -3 + 1
y = -2

y = x + 1
when x = -1
y = -1 + 1
y = 0

y = x + 1
when x = 0
y = 0 + 1
y = 1

y = x + 1
when x = 2
y = 2 + 1
y = 3

(4 marks)

(b) On the graph paper on the next page, draw the graph of

y = x + 1 using the table above. Use a scale of 2 cm = 1 unit on the x-axis and 2 cm = 1 unit on the y-axis.

(6 marks)

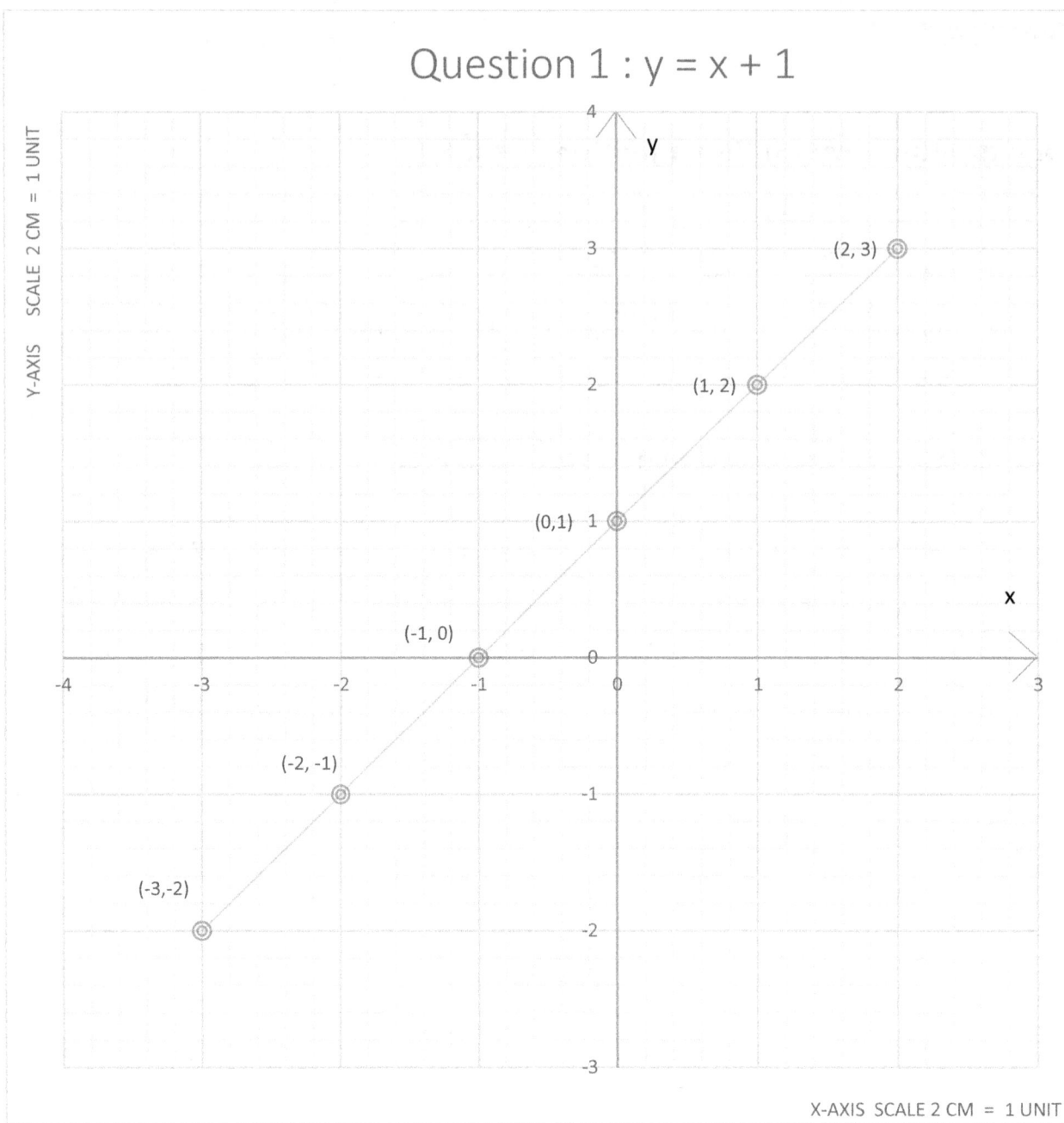

Question 2

Consider the straight line equation: y = x + 2.

x	-3	-2	-1	0	1	2
y	-1	0	1	2	3	4

(a) Complete the table above for: y = x + 2.

y = x + 2   y = x + 2   y = x + 2   y = x + 2
when x = -2   when x = -1   when x = 0   when x = 1
y = -2 + 2   y = -1 + 2   y = 0 + 2   y = 1 + 2
y = 0   **y = 1**   **y = 2**   **y = 3**

(4 marks)

(b) On the graph paper on the next page, draw the graph of

y = x + 2 using the table above. Use an appropriate scale.

(6 marks)

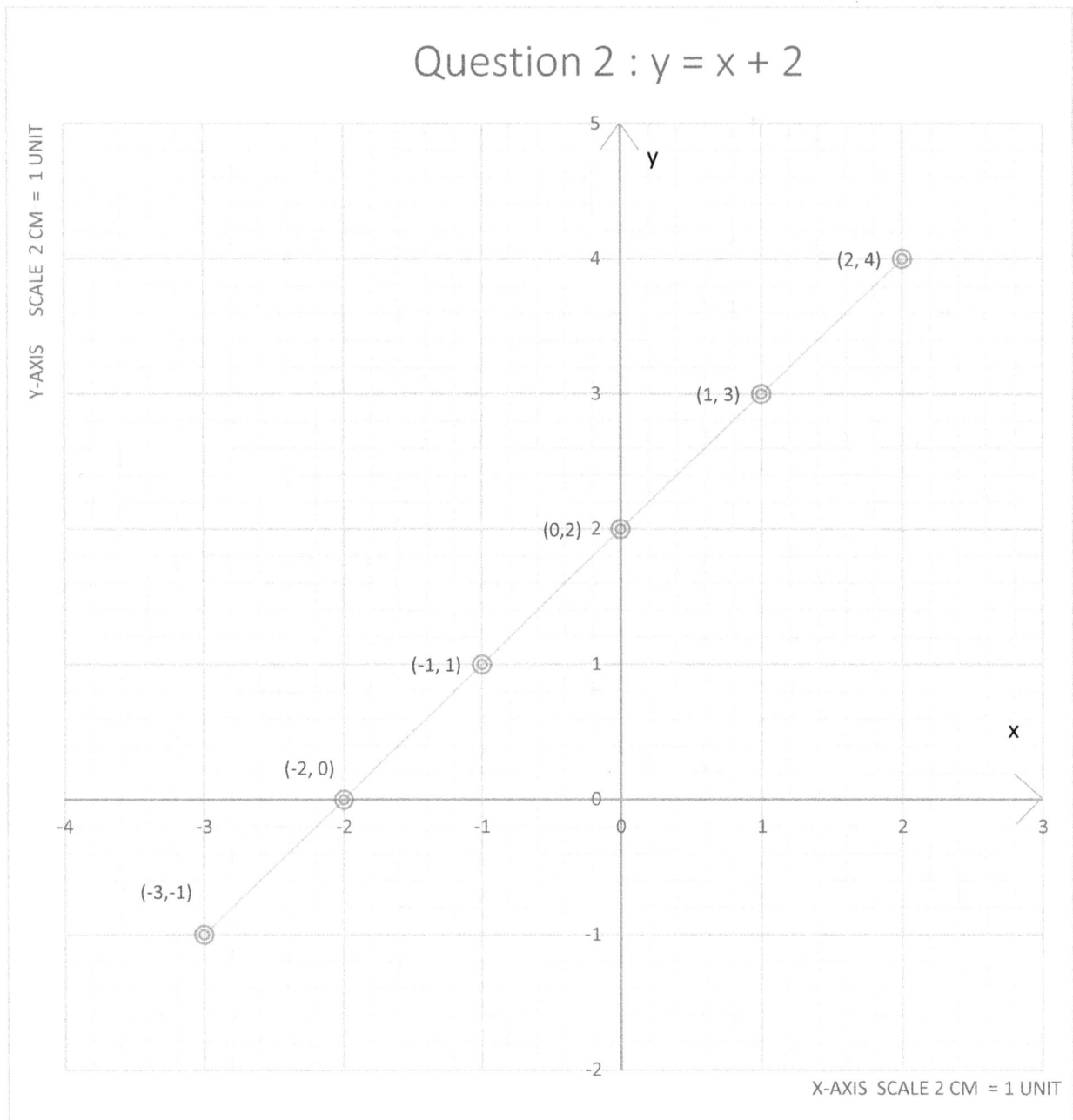

Question 3

Consider the straight line equation: y = x - 3.

x	-3	-2	-1	0	1	2	3
y	-6	-5	-4	-3	-2	-1	0

(a) Complete the table above for: y = x - 3.

y = x - 3   y = x - 3   y = x - 3   y = x - 3   y = x - 3
when x = -2   when x = -1   when x = 0   when x = 1   when x = 3
y = -2 - 3   y = -1 - 3   y = 0 - 3   y = 1 - 3   y = 3 - 3
y = -5   **y = -4**   **y = -3**   **y = -2**   **y = 0**

(4 marks)

(b) On the graph paper on the next page, draw the graph of

y = x - 3 using the table above. Use an appropriate scale.

(6 marks)

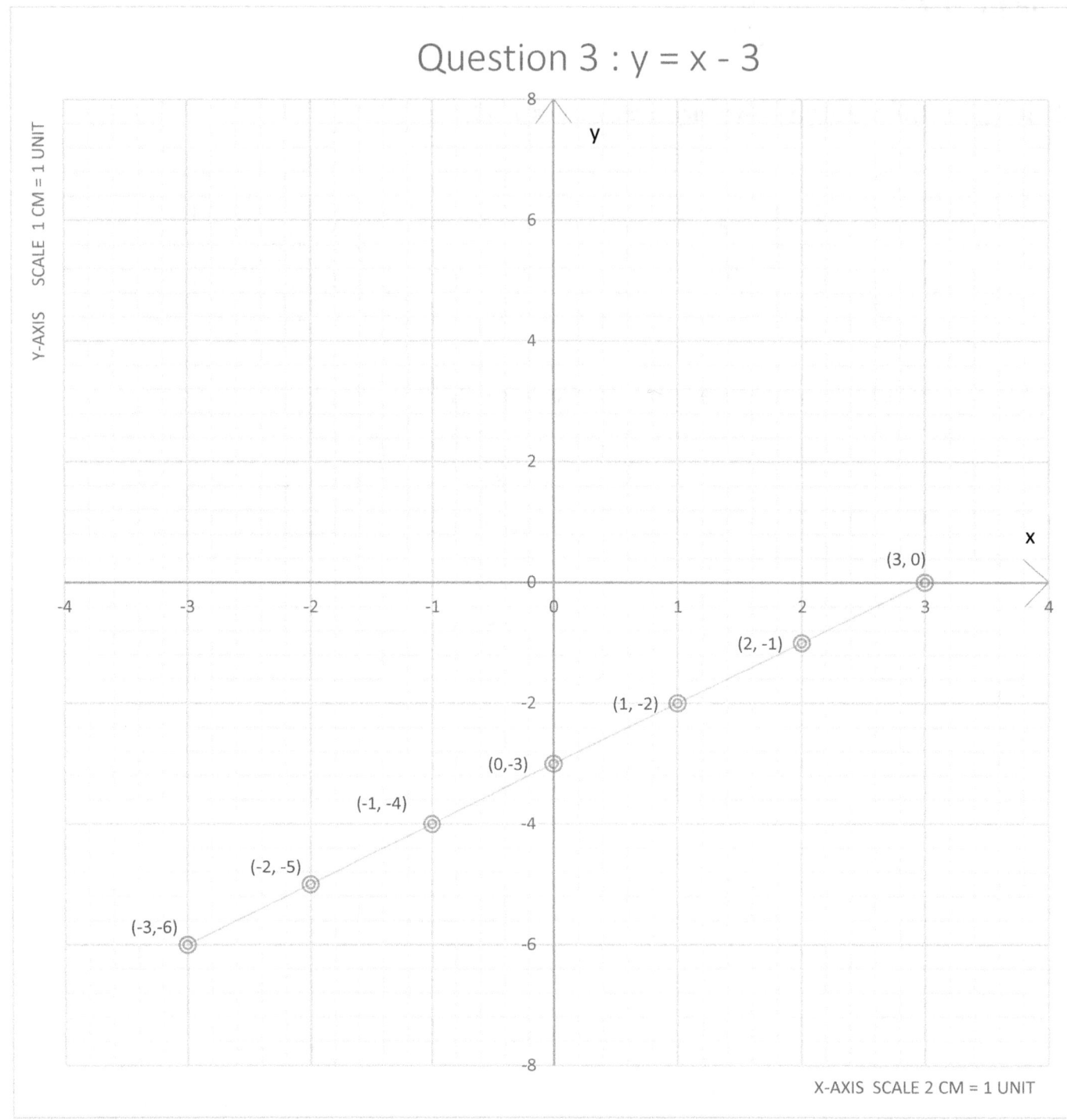

Question 4

Consider the straight line equation: y = 2x + 4.

x	-3	-2	-1	0	1	2	3
y	-2	0	2	4	6	8	10

(a) Complete the table above for: y = 2x + 4.

y = 2x + 4
when x = -2
y = 2(-2) + 4
y = -4 + 4
y = 0

y = 2x + 4
when x = -1
y = 2(-1) + 4
y = -2 + 4
y = 2

y = 2x + 4
when x = 0
y = 2(0) + 4
y = + 4
y = 4

y = 2x + 4
when x = 1
y = 2(1) + 4
y = 2 + 4
y = 6

y = 2x + 4
when x = 2
y = 2(2) + 4
y = 4 + 4
y = 8

(5 marks)

(b) On the graph paper on the next page, draw the graph of

y = 2x + 4 using the table above. Use an appropriate scale.

(6 marks)

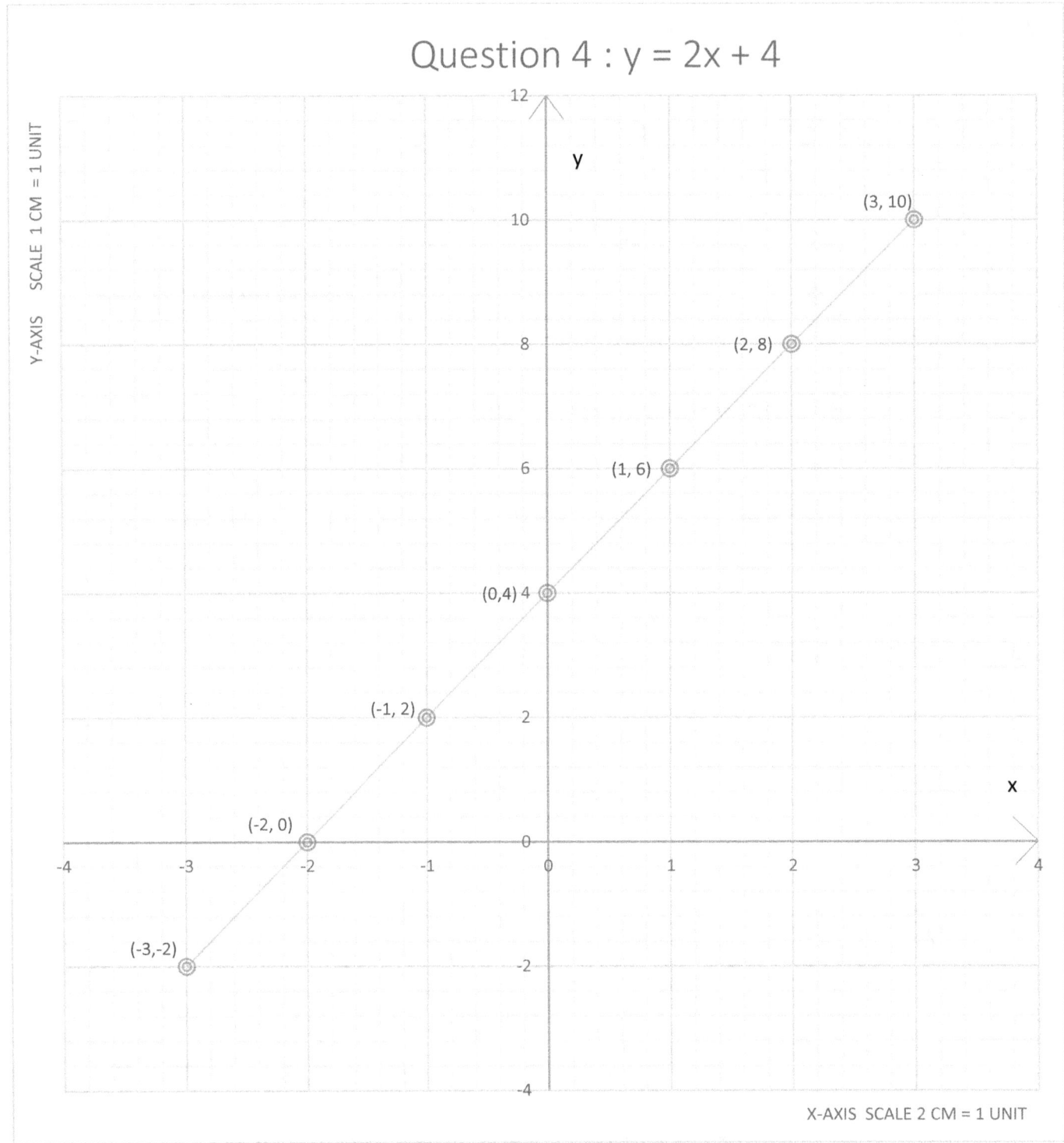

Question 5

Consider the straight line equation: y = 2x - 3.

x	-3	-2	-1	0	1	2	3
y	-9	-7	-5	-3	-1	1	3

(a) Complete the table above for: y = 2x - 3.

y = 2x - 3	y = 2x - 3	y = 2x - 3	y = 2x - 3	y = 2x - 3	y = 2x - 3	y = 2x - 3
when x = -3	when x = -2	when x = -1	when x = 0	when x = 1	when x = 2	when x = 3
y = 2(-3) - 3	y = 2(-2) - 3	y = 2(-1) - 3	y = 2(0) - 3	y = 2(1) - 3	y = 2(2) - 3	y = 2(3) - 3
y = -6 - 3	y = -4 - 3	y = -2 - 3	y = 0 - 3	y = 2 - 3	y = 4 - 3	y = 6 - 3
y = -9	**y = -7**	**y = -5**	**y = -3**	**y = -1**	**y = 1**	**y = 3**

(7 marks)

(b) On the graph paper on the next page, draw the graph of y = 2x - 3 using the table above. Use an appropriate scale.

(6 marks)

Complete the following statements.

(c) The gradient of the straight line y = 2x - 3 is : _____2_____

(d) The y intercept of the straight line y = 2x - 3 is _____-3_____

(2 marks)

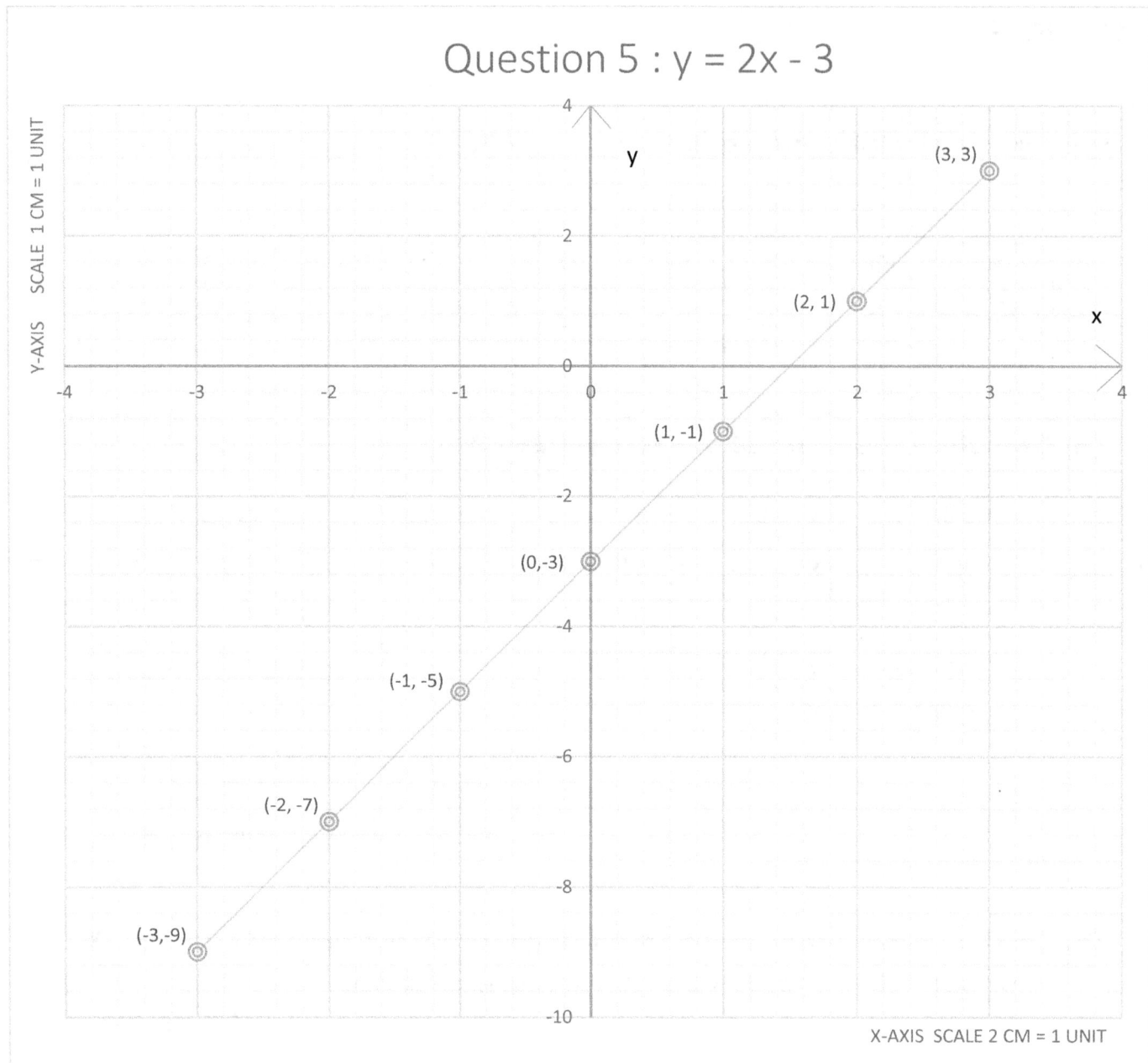

Question 6

Consider the straight line equation: $y = \frac{1}{2}x + 3$

x	-3	-2	-1	0	1	2	3	4	5	6
y	1.5	2	2.5	3	3.5	4	4.5	5	5.5	6

(a) Complete the table above for: $y = \frac{1}{2}x + 3$.

$y = \frac{1}{2}x + 3$
when x = -3
$y = \frac{1}{2}(-3) + 3$
$y = -1.5 + 3$
y = 1.5

$y = \frac{1}{2}x + 3$
when x = -2
$y = \frac{1}{2}(-2) + 3$
$y = -1 + 3$
y = 2

$y = \frac{1}{2}x + 3$
when x = -1
$y = \frac{1}{2}(-1) + 3$
$y = -0.5 + 3$
y = 2.5

$y = \frac{1}{2}x + 3$
when x = 0
$y = \frac{1}{2}(0) + 3$
$y = 0 + 3$
y = 3

$y = \frac{1}{2}x + 3$
when x = 1
$y = \frac{1}{2}(1) + 3$
$y = 0.5 + 3$
y = 3.5

$y = \frac{1}{2}x + 3$
when x = 2
$y = \frac{1}{2}(2) + 3$
$y = 1 + 3$
y = 4

$y = \frac{1}{2}x + 3$
when x = 3
$y = \frac{1}{2}(3) + 3$
$y = 1.5 + 3$
y = 4.5

$y = \frac{1}{2}x + 3$
when x = 4
$y = \frac{1}{2}(4) + 3$
$y = 2 + 3$
y = 5

$y = \frac{1}{2}x + 3$
when x = 5
$y = \frac{1}{2}(5) + 3$
$y = 2.5 + 3$
y = 5.5

$y = \frac{1}{2}x + 3$
when x = 6
$y = \frac{1}{2}(6) + 3$
$y = 3 + 3$
y = 6

(10 marks)

(b) On the graph paper on the next page, draw the graph of $y = \frac{1}{2}x + 3$ using the table above. Use an appropriate scale.

(6 marks)

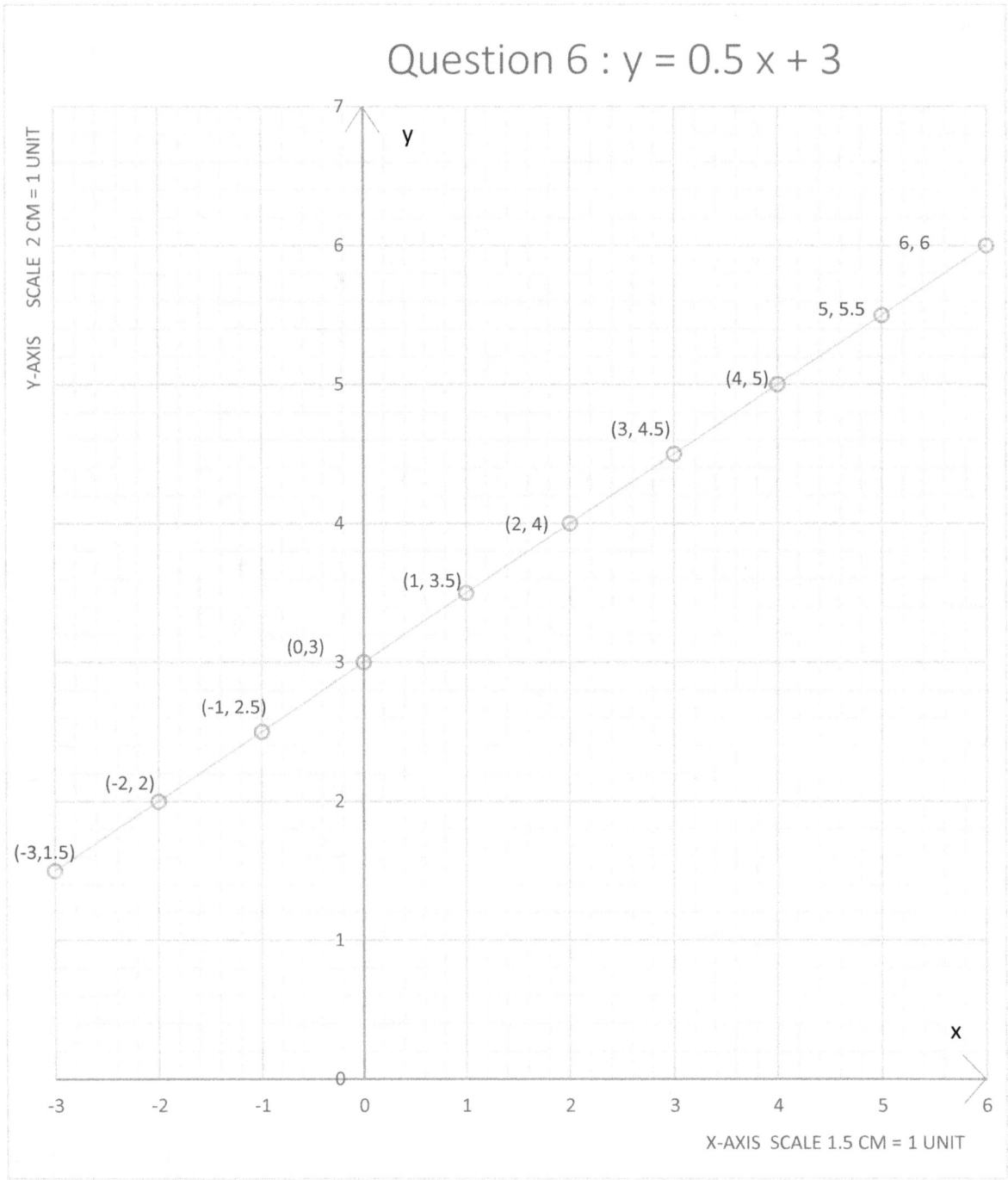

Question 7

Consider the straight line equation: $y = -\frac{1}{2}x + 3$

x	-3	-2	-1	0	1	2	3	4
y	4.5	4	3.5	3	2.5	2	1.5	1

(a) Complete the table above for: $y = -\frac{1}{2}x + 3$

$y = -\frac{1}{2}x + 3$
when x = -3
$y = -\frac{1}{2}(-3) + 3$
$y = 1.5 + 3$
y = 4.5

$y = -\frac{1}{2}x + 3$
when x = -2
$y = -\frac{1}{2}(-2) + 3$
$y = 1 + 3$
y = 4

$y = -\frac{1}{2}x + 3$
when x = -1
$y = -\frac{1}{2}(-1) + 3$
$y = 0.5 + 3$
y = 3.5

$y = -\frac{1}{2}x + 3$
when x = 0
$y = -\frac{1}{2}(0) + 3$
$y = 0 + 3$
y = 3

$y = -\frac{1}{2}x + 3$
when x = 1
$y = -\frac{1}{2}(1) + 3$
$y = -0.5 + 3$
y = 2.5

$y = -\frac{1}{2}x + 3$
when x = 2
$y = -\frac{1}{2}(2) + 3$
$y = -1 + 3$
y = 2

$y = -\frac{1}{2}x + 3$
when x = 3
$y = -\frac{1}{2}(3) + 3$
$y = -1.5 + 3$
y = 1.5

$y = -\frac{1}{2}x + 3$
when x = 4
$y = -\frac{1}{2}(4) + 3$
$y = -2 + 3$
y = 1

(8 marks)

(b) On the graph paper on the next page, draw the graph of $y = -\frac{1}{2}x + 3$ using the table above. Use an appropriate scale.

(6 marks)

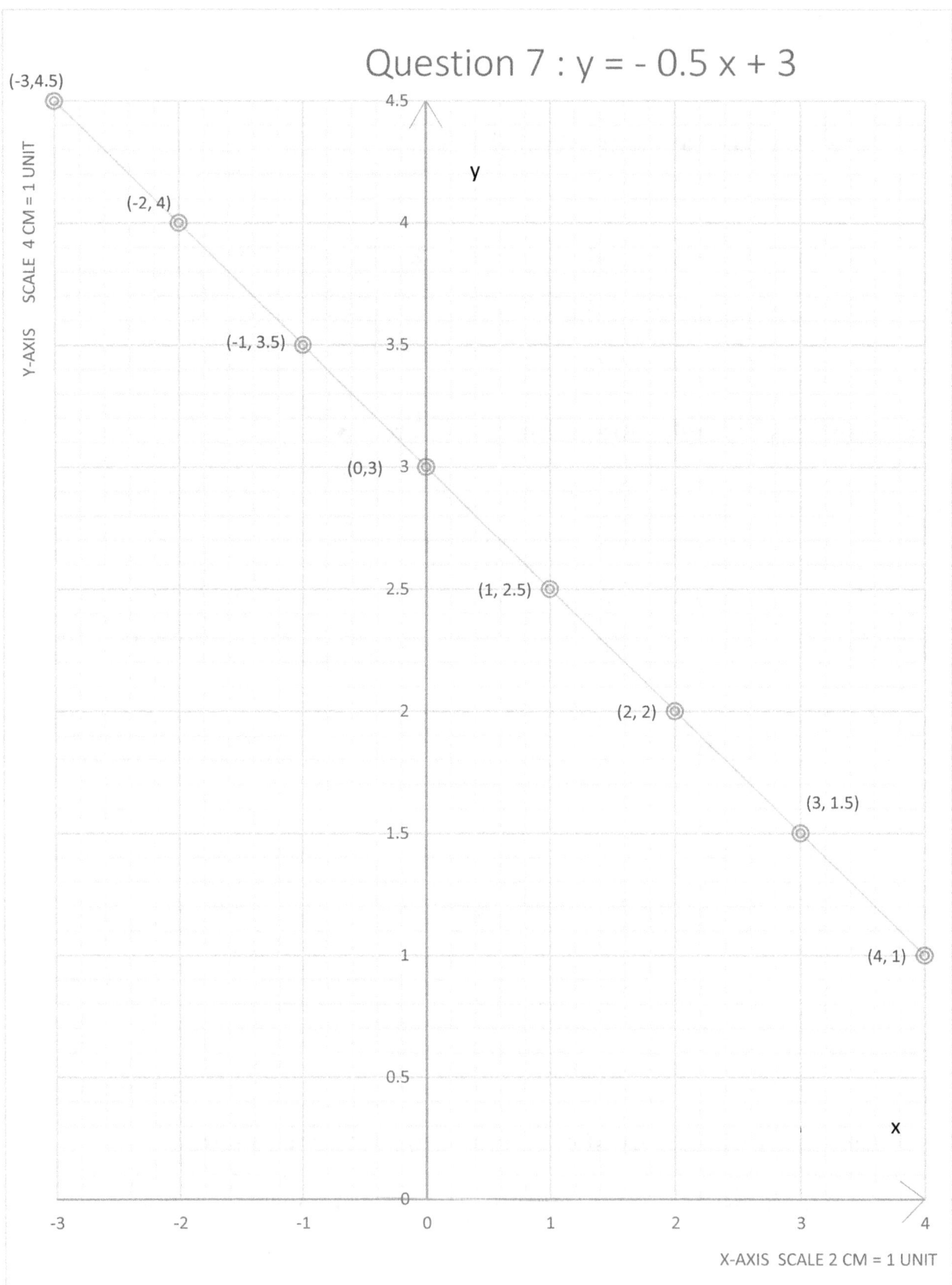

Question 8

Determine whether the following pairs of lines are perpendicular or parallel to one another. Give an explanation in each case.

(a) y = 2x + 3 and y = 2x - 3

The lines y = 2x + 3 and y = 2x − 3 are parallel because they have the same gradient, m = 2.

(b) y = 3x + 8 and y = $-\frac{1}{3}$x + 4

The lines y = 3x + 8 and y = $-\frac{1}{3}$x + 4 are perpendicular because the product of their gradient is = -1.

(c) 2y = 10x + 1 and 3y = 15x + 3

$$2y = 10x + 1 \qquad\qquad 3y = 15x + 3$$

$$y = \frac{10x + 1}{2} \qquad\qquad y = \frac{15x + 3}{3}$$

$$y = 5x + \frac{1}{2} \qquad\qquad y = 5x + 1$$

The lines are parallel because the gradient is the same, m = 5.

(d) 6y = 1x + 4 and 2y = -12x + 8

$$6y = 1x + 4 \qquad\qquad 2y = -12x + 8$$

$$y = \frac{1x + 4}{6} \qquad\qquad y = \frac{-12x + 8}{2}$$

$$y = \frac{1}{6}x + \frac{4}{6} \qquad\qquad y = -6x + 4$$

$$y = \frac{1}{6}x + \frac{2}{3}$$

The lines are perpendicular because the product of their gradient is = -1.

$$\frac{1}{6} \times -6 = -1$$

(8 marks)

END OF WORKSHEET

868

868TUTORS

TUTORS

Preparation for

High School Mathematics

Subject of the Formula

Solutions

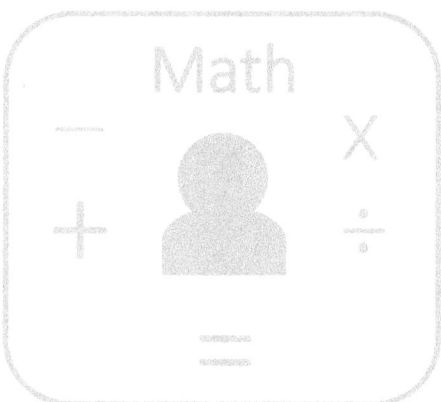

Instructions and Tips:

- ✓ You have 40 minutes to complete this worksheet
- ✓ This worksheet consists of 8 questions
- ✓ Write answers in the spaces provided
- ✓ All working must be clearly shown

Student Name: _____

Student ID: _____

Date: __/__/____

Total Score:

Highest Score:

Tutor's Comments:

Access more free worksheets at www.868tutors.com

Question 1

Make y the subject of the formula in the following formulae:

(a)   **a = y + b + c**

y + b + c = a

y + b = a - c

$$\boxed{y = a - c - b}$$

(b)   **b = y - a**

y – a = b

$$\boxed{y = b + a}$$

(c)   **C = h – y - z**

h – y – z = C

-y + h – z = C

-y + h = C + z

-y = C + z – h

$$\boxed{y = -C - z + h}$$

(d)   **V = -y**

-y = V

$$\boxed{y = -V}$$

(4 marks)

Question 2

Make a the subject of the formula in the following formulae:

(a) H = a + b + c

a + b + c = h

$$\boxed{a = h - b - c}$$

(b) b = y − 3a

y - 3a = b

-3a = b - y

$$\boxed{a = \frac{b - y}{-3}}$$

(c) C = h − y − z + 2a

h − y − z + 2a = C

2a = C − h + y + z

$$\boxed{a = \frac{C - h + y + z}{2}}$$

(d) V = −y − a

-y - a = V

-a = V + y

$$\boxed{a = -V - y}$$

(4 marks)

Question 3

Consider the formula $V = \pi r^2 h$ for the volume of a cylinder.

(a) Make h the subject of the formula.

$V = \pi r^2 h$

$\pi r^2 h = V$

$\boxed{h = \dfrac{V}{\pi r^2}}$

(1 mark)

(b) Make r the subject of the formula.

$V = \pi r^2 h$

$\pi r^2 h = V$

$r^2 = \dfrac{V}{\pi h}$ $\quad \boxed{r = \left(\dfrac{V}{\pi h}\right)^{0.5}}$

(1 mark)

Consider the formula $C = 2\pi r$ for the circumference of a circle.

(c) Make r the subject of the formula.

$C = 2\pi r$

$2\pi r = C$

$\boxed{r = \dfrac{C}{2\pi}}$

(1 mark)

Consider the formula $A = \pi r^2$ for the area of a circle.

(d) Make r the subject of the formula.

$\pi r^2 = A$

$r^2 = \dfrac{A}{\pi}$ $\quad \boxed{r = \left(\dfrac{A}{\pi}\right)^{0.5}}$

(1 mark)

Question 4

Consider the formula $V = \frac{4}{3}\pi r^3$ for the volume of a sphere.

(a) Make r the subject of the formula.

$V = \frac{4}{3}\pi r^3 \quad \frac{4}{3}\pi r^3 = V \quad r^3 = V \div \frac{4\pi}{3} \quad r^3 = V \times \frac{3}{4\pi}$

$r^3 = \frac{3V}{4\pi} \quad \boxed{r = \left(\frac{3V}{4\pi}\right)^{\frac{1}{3}}}$

(1 mark)

Consider the formula $V = 4\pi r^2$ for the surface area of a sphere.

(b) Make r the subject of the formula.

$V = 4\pi r^2$

$4\pi r^2 = V$

$r^2 = \frac{V}{4\pi}$

$\boxed{r = \left(\frac{V}{4\pi}\right)^{\frac{1}{2}}}$

(1 mark)

Consider the formula $A = s^2$ for the area of a square.

(c) Make s the subject of the formula.

$A = s^2$

$s^2 = A$

$\boxed{s = \sqrt{A}}$

(1 mark)

Question 5

Consider the formula $c^2 = a^2 + b^2$

(a) Make a the subject of the formula.

$a^2 + b^2 = c^2$

$a^2 = c^2 - b^2$

$\boxed{a = \sqrt{c^2 - b^2}}$

(1 mark)

(b) Make b the subject of the formula.

$c^2 = a^2 + b^2$

$a^2 + b^2 = c^2$

$b^2 = c^2 - a^2$

$\boxed{b = \sqrt{c^2 - a^2}}$

(1 mark)

(d) Make c the subject of the formula.

$c^2 = a^2 + b^2$

$\boxed{c = \sqrt{a^2 + b^2}}$

(1 mark)

Question 6

Consider the formula for the volume of a cube $V = s^3$

(a) Make s the subject of the formula.

$V = s^3 \quad s^3 = V \quad \boxed{s = \sqrt[3]{V}}$

(1 mark)

Consider the formula for the area of a parallelogram $A = bh$

(b) Make b the subject of the formula.

$A = bh \quad b \times h = A \quad \boxed{b = \dfrac{A}{h}}$

(1 mark)

Consider the formula for the volume of a cuboid $V = abc$

(c) Make a the subject of the formula.

$V = abc \quad a \times b \times c = V \quad \boxed{a = \dfrac{V}{bc}}$

(d) Make b the subject of the formula.

$V = abc \quad a \times b \times c = V \quad \boxed{b = \dfrac{V}{ac}}$

(e) Make c the subject of the formula.

$V = abc \quad a \times b \times c = V \quad \boxed{c = \dfrac{V}{ab}}$

(3 marks)

Question 7

Consider the formula for the area of a trapezium: $A = \dfrac{h}{2}(a+b)$

(a) Make h the subject of the formula.

$A = \dfrac{h}{2}(a+b) \quad A \div (a+b) = \dfrac{h}{2} \quad A \times \dfrac{1}{(a+b)} = \dfrac{h}{2} \quad \dfrac{A}{(a+b)} = \dfrac{h}{2}$ (Cross-multiplying)

$h(a+b) = 2A \quad \boxed{h = \dfrac{2A}{(a+b)}}$

(1 mark)

(b) Make a the subject of the formula.

$A = \dfrac{h}{2}(a+b) \quad A \div \dfrac{h}{2} = (a+b) \quad A \times \dfrac{2}{h} = (a+b)$

$\dfrac{2A}{h} = (a+b) \quad (a+b) = \dfrac{2A}{h} \quad \boxed{a = \dfrac{2A}{h} - b}$

(1 mark)

(c) Make b the subject of the formula.

$A = \dfrac{h}{2}(a+b) \quad A \div \dfrac{h}{2} = (a+b) \quad A \times \dfrac{2}{h} = (a+b)$

$\dfrac{2A}{h} = (a+b) \quad (a+b) = \dfrac{2A}{h} \quad \boxed{b = \dfrac{2A}{h} - a}$

(1 mark)

Question 8

Make the letter indicated in brackets the subject of the formula of the following formulae:

(a) $y = qa^4$ [a]

$qa^4 = y$

$a^4 = \dfrac{y}{q}$

$$\boxed{a = \left(\dfrac{y}{q}\right)^{\frac{1}{4}}}$$

(b) $g = u^2 + vz$ [z]

$u^2 + vz = g$

$vz = g - u^2$

$$\boxed{z = \dfrac{g - u^2}{v}}$$

(c) $M^2 = fzs$ [M]

$M^2 = fzs$

$$\boxed{M = \sqrt{fzs}}$$

(d) $k = ft^6$ [t]

$ft^6 = k$

$t^6 = \dfrac{k}{f}$

$$\boxed{t = \left(\dfrac{k}{f}\right)^{\frac{1}{6}}}$$

(4 marks)

END OF WORKSHEET

868 TUTORS

Preparation for

High School Mathematics

Trigonometry

(Right angled triangles)

Solutions

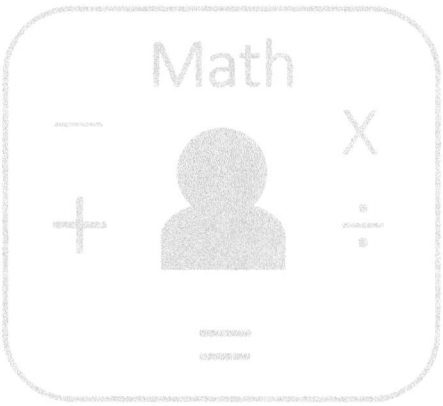

Instructions and Tips:

- ✓ You have 75 minutes to complete this worksheet
- ✓ This worksheet consists of 6 questions
- ✓ Write answers in the spaces provided
- ✓ All working must be clearly shown

Student Name: _____

Student ID: _____

Date: __/__/____

Total Score:

Highest Score:

Tutor's Comments:

Access more free worksheets at www.868tutors.com

Question 1

Each right-angled triangle below has an 'angle of interest' indicated. This 'angle of interest' is an angle that you can be asked to solve. Using the information below, label the right angle, the 'angle of interest' (as x), the opposite, the adjacent and the hypotenuse sides for the following triangles.

(a)

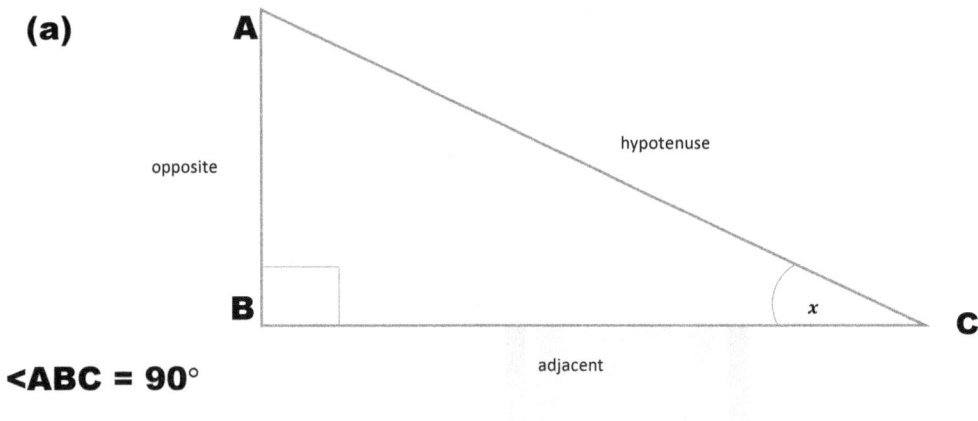

<ABC = 90°

<ACB = angle of interest

(5 marks)

(b)

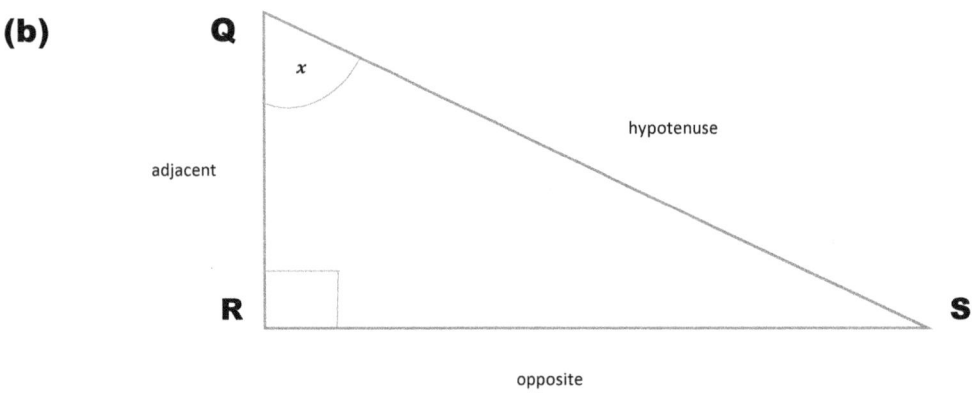

<QRS = 90°

<RQS = angle of interest

(5 marks)

(c)

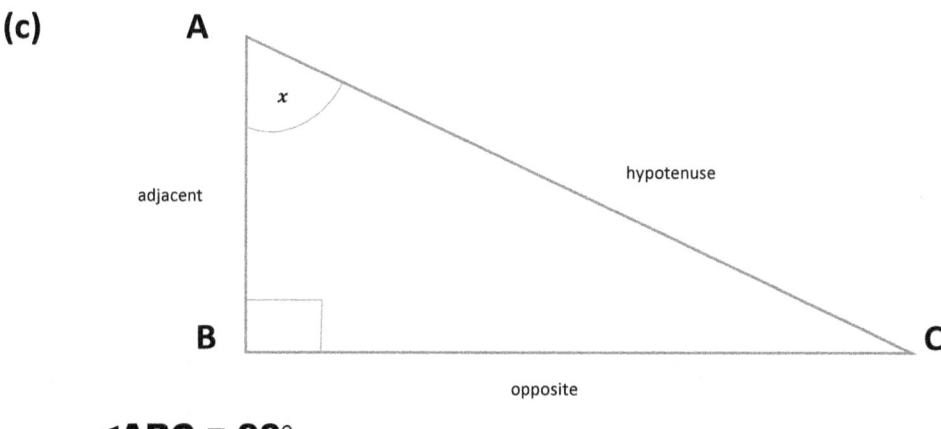

<ABC = 90°
<BAC = angle of interest

(5 marks)

(d)

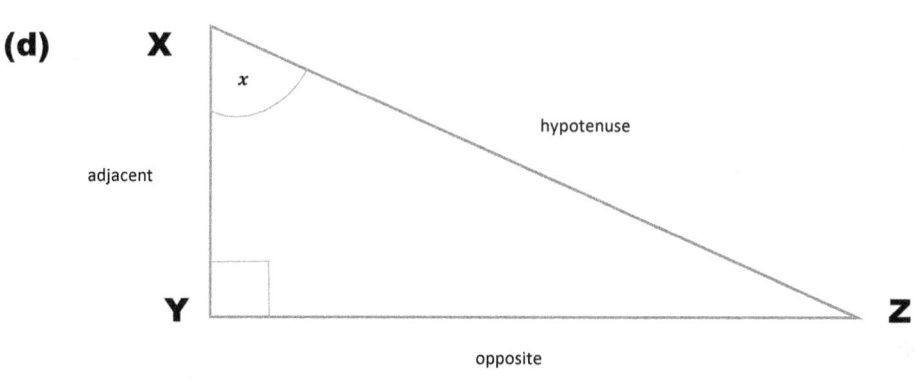

<XYZ = 90°
<YXZ = angle of interest

(5 marks)

(e)

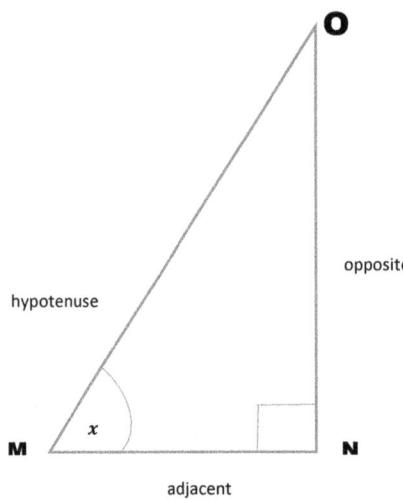

<MNO=90°

<OMN=angle of interest

(5 marks)

(f)

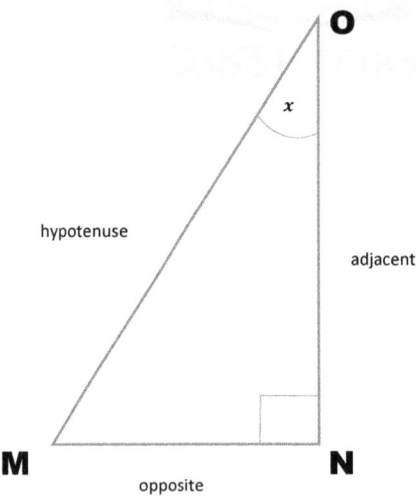

<MNO=90°

<MON=angle of interest

(5 marks)

Question 2

Pythagoras is credited with a theorem involving right angled triangles. Apply the Pythagoras theorem in the following questions to find the lengths of the unknown sides.

(a) <ABC=90°

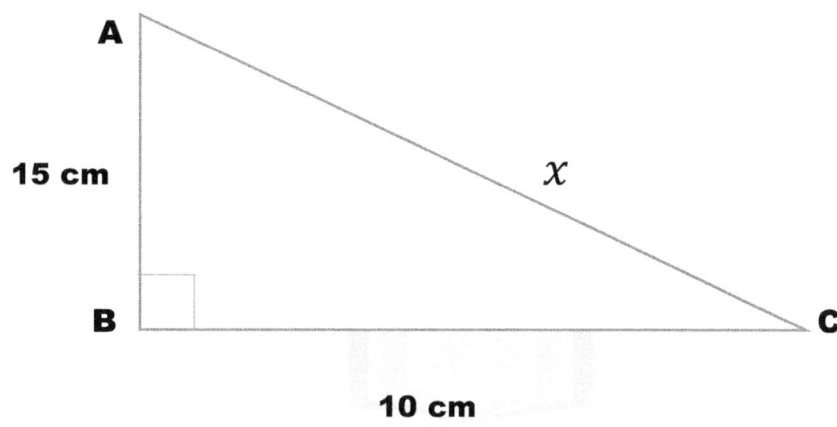

$(AB)^2 + (BC)^2 = (AC)^2$

$(15)^2 + (10)^2 = (AC)^2$

$225 + 100 = (AC)^2$

$(AC)^2 = 225 + 100$

$(AC)^2 = 325$

$\boxed{AC = 18.03 \text{ cm (to 2 decimal places)}}$

(3 marks)

(b) <QRS=90°

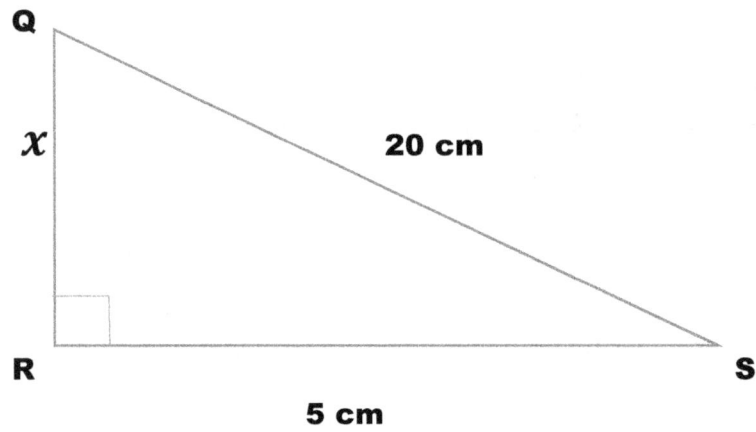

$(QR)^2 + (RS)^2 = (QS)^2$

$(QR)^2 = (QS)^2 - (RS)^2$

$(QR)^2 = (20)^2 - (5)^2$

$(QR)^2 = 400 - 25$

$(QR)^2 = 375$

QR = 19.36 cm (to 2 decimal places)

(3 marks)

(c) <QRS=90°

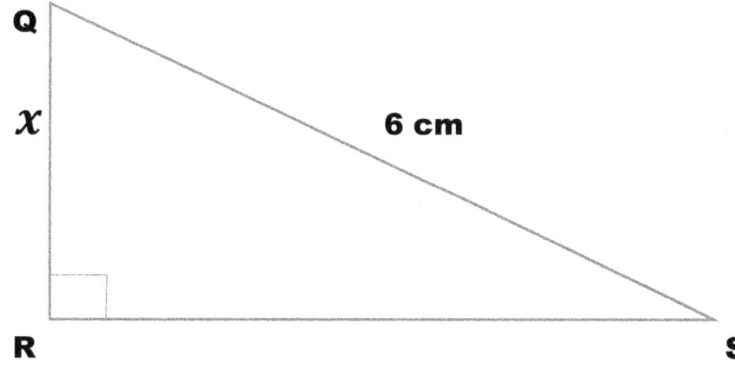

$(QR)^2 + (RS)^2 = (QS)^2$

$(QR)^2 = (QS)^2 - (RS)^2$

$(QR)^2 = (6)^2 - (4)^2$

$(QR)^2 = 36 - 16$

$(QR)^2 = 20$

QR = 4.47 cm (to 2 decimal places)

(3 marks)

(d) <QRS=90°

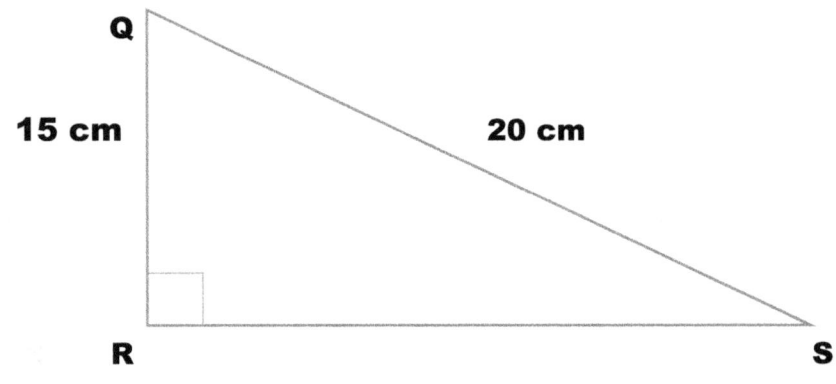

$(QR)^2 + (RS)^2 = (QS)^2$

$(RS)^2 = (QS)^2 - (QR)^2$

$(RS)^2 = (20)^2 - (15)^2$

$(RS)^2 = 400 - 225$

$(RS)^2 = 175$

RS = 13.23 cm (to 2 decimal places)

(3 marks)

Question 3 (Angle of Elevation and Angle of Depression)

For each question below, draw the appropriate right angled triangle.

Junior Addams is 1.8m tall. He stands 20 meters away from his local television antenna. The antenna is 25 meters high. What is the angle of elevation of the top of the antenna from his eyes?

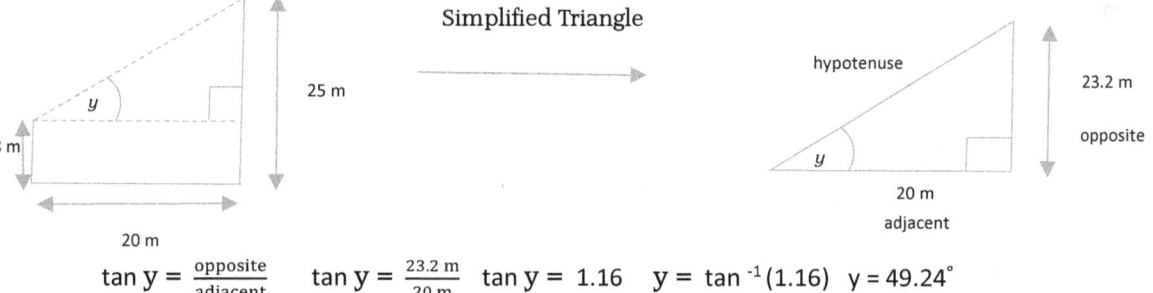

$\tan y = \dfrac{\text{opposite}}{\text{adjacent}}$ $\tan y = \dfrac{23.2 \text{ m}}{20 \text{ m}}$ $\tan y = 1.16$ $y = \tan^{-1}(1.16)$ $y = 49.24°$

angle of elevation of the top of the antenna from his eyes = 49.24° (to 2 decimal places)

(3 marks)

Question 4

A farmer in Los Iros enjoys the scenic view from the top of a cliff that is 40 m high. The farmer is 1.8 m tall and he is standing. He observes a personal watercraft that has stalled in the ocean. The angle of depression is 20°. If he is line with the watercraft, calculate the distance between the watercraft and the farmer.

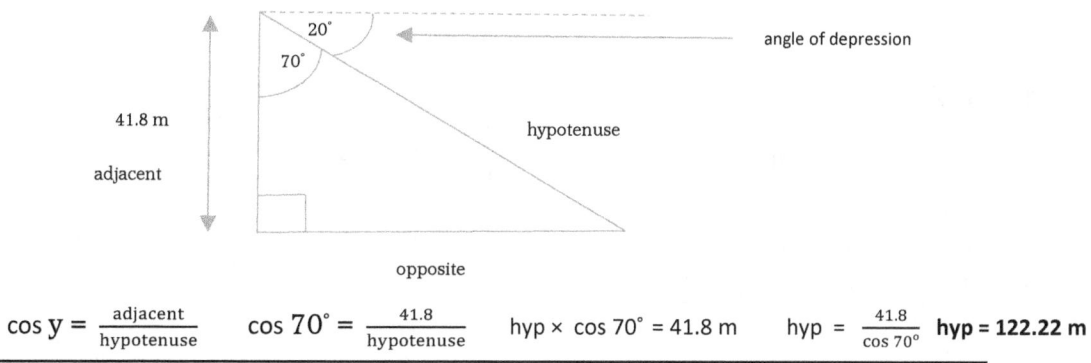

$\cos y = \dfrac{\text{adjacent}}{\text{hypotenuse}}$ $\cos 70° = \dfrac{41.8}{\text{hypotenuse}}$ $\text{hyp} \times \cos 70° = 41.8 \text{ m}$ $\text{hyp} = \dfrac{41.8}{\cos 70°}$ $\text{hyp} = 122.22 \text{ m}$

distance between the watercraft and farmer = 122.22 m (to 2 decimal places)

(3 marks)

Question 5

The diagram below is a simplified diagram of a possible scenario. In the diagram, AB represents a radio tower that is vertical. The radio tower rests on a horizontal plane and A, D and C are points on the horizontal plane.

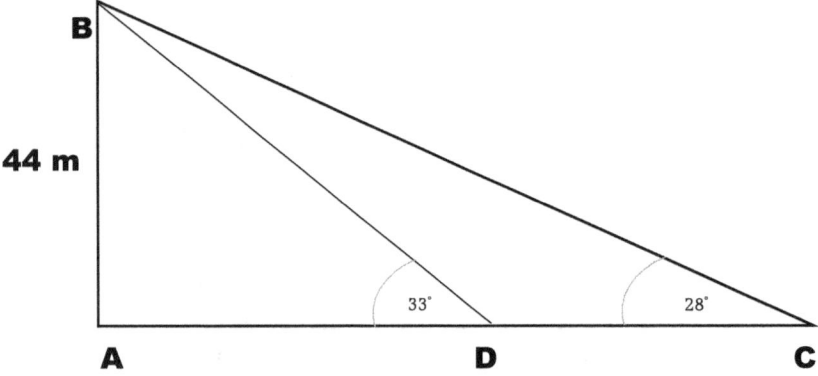

AB = 44 m

The angles of elevation of the top of the tower B from D and C are 33° and 28° respectively

(a) Indicate on the diagram above the relevant angles of elevation

(2 marks)

Determine

(b) the length of AD

let angle ADB = y $\tan y = \frac{\text{opposite}}{\text{adjacent}}$ $\tan y = \frac{44 \text{ m}}{\text{adjacent}}$ $\tan 33° = \frac{44 \text{ m}}{\text{adjacent}}$ $\text{adj} \times \tan 33° = 44\text{m}$ $\text{adj} = \frac{44 \text{ m}}{\tan 33°}$

adj = 67.75 m

length of AD = 67.75 m (to 2 decimal places)

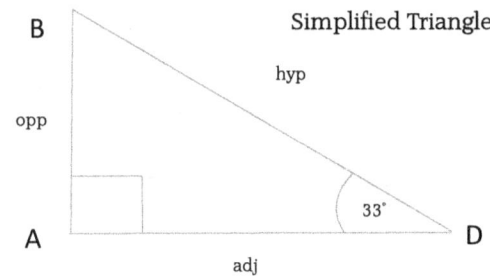

Simplified Triangle

(3 marks)

(c) the length of DC

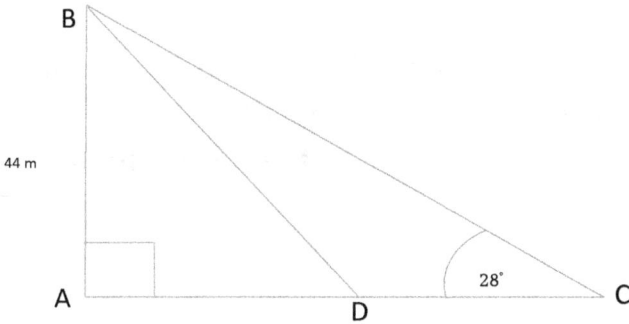

Length of DC = Length of AC − Length of AD

Firstly, Determine the length of AC

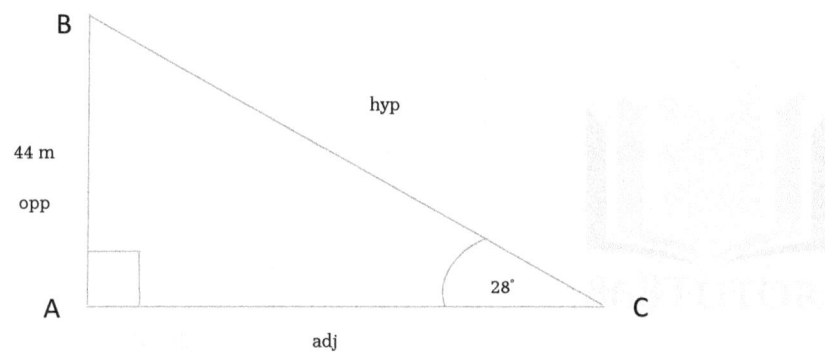

Let angle ACB = y

$\tan y = \dfrac{\text{opposite}}{\text{adjacent}}$

$\tan 28° = \dfrac{44 \text{ m}}{\text{adj}}$

adj × tan 28° = 44m

$\text{adj} = \dfrac{44 \text{ m}}{\tan 28°}$

$\text{adj} = \dfrac{44 \text{ m}}{\tan 28°}$ adj = 82.75 m   AC = 82.75 m

Length of DC = Length of AC − Length of AD

Recall length of AD = 67.75 m

length of DC = 82.75 m - 67.75 m

Length of DC = 15.00 m (to 2 decimal places)   (3 marks)

Question 6

The diagram below is a simplified diagram of a possible scenario. In the diagram RS represents an oil rig. Assume that the oil rig is vertical. The oil rig rests on a horizontal plane and P, Q and R are points on the horizontal plane.

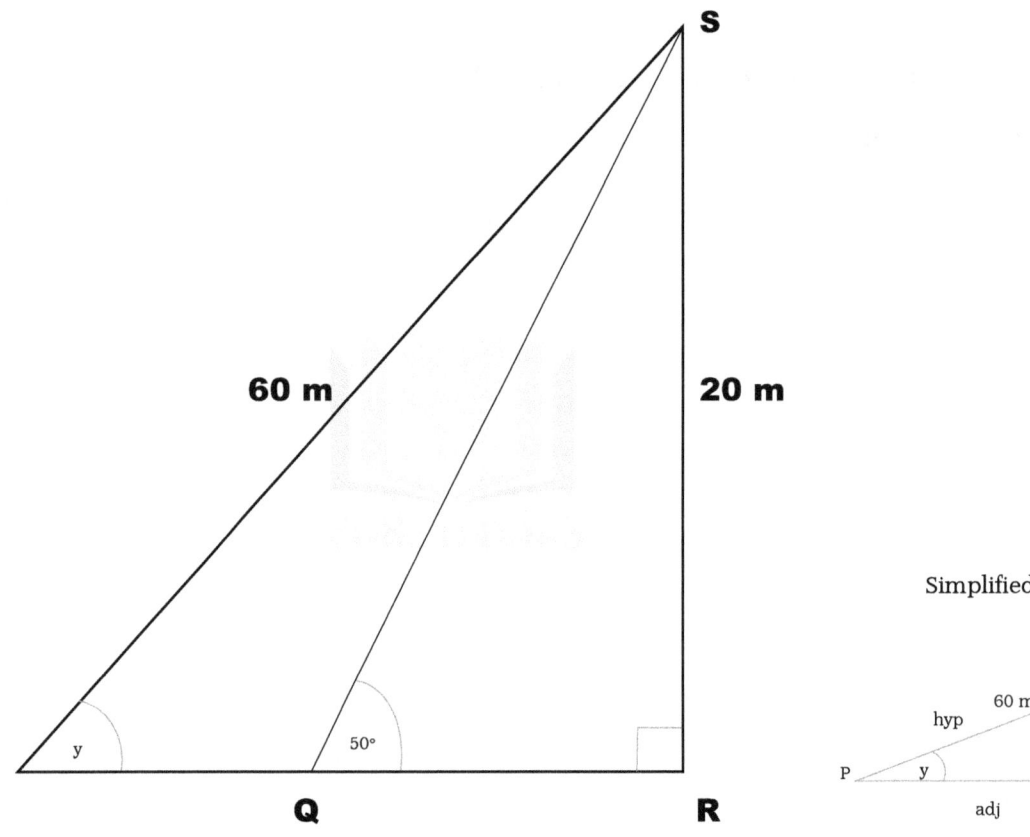

Simplified Triangle

∠SQR = 50°, ∠PRS = 90°, PS = 60 m, RS = 20 m

Determine, giving your answer to 2 decimal places

(a) the size of ∠SPR

$\sin y = \frac{\text{opposite}}{\text{hypotenuse}}$ $\sin y = \frac{20 \text{ m}}{60 \text{ m}}$ $\sin y = \frac{1}{3}$ $y = \sin^{-1}\left(\frac{1}{3}\right)$ $\boxed{y = 19.47° \text{ (to 2 decimal places)}}$

(2 marks)

(b) the length of PQ, in cm

length of PQ = length of PR − length of QR Determining the length of PR (simplified triangle)

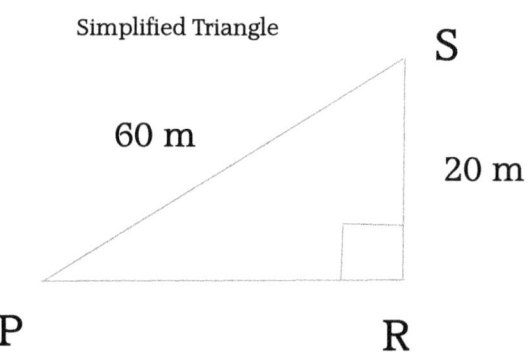

$(PR)^2 + (RS)^2 = (PS)^2$

$(PR)^2 = (PS)^2 − (RS)^2$

$(PR)^2 = (60)^2 − (20)^2$

$(PR)^2 = 3600 − 400$

$(PR)^2 = 3200$

PR = 56.57 m

Determining the length of QR (simplified triangle)

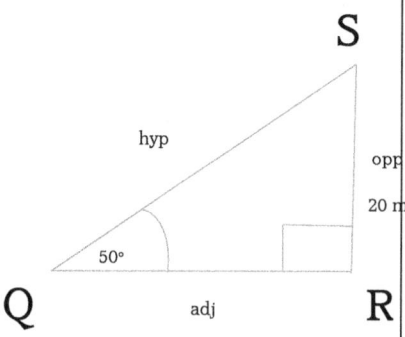

let angle RQS = y

$\tan y = \frac{\text{opposite}}{\text{adjacent}}$

$\tan 50° = \frac{20}{\text{adjacent}}$

20 = adj × tan 50°

adj = $\frac{20}{\tan 50°}$ adj = 16.78 m   QR = 16.78 m

length of PQ = length of PR − length of QR

length of PQ = 56.57 m − 16.78 m

length of PQ = 39.79 m (to 2 decimal places)

(5 marks)

END OF WORKSHEET

868 TUTORS

Preparation for

High School Mathematics

Trigonometry

(Non-right angled triangles)

Solutions

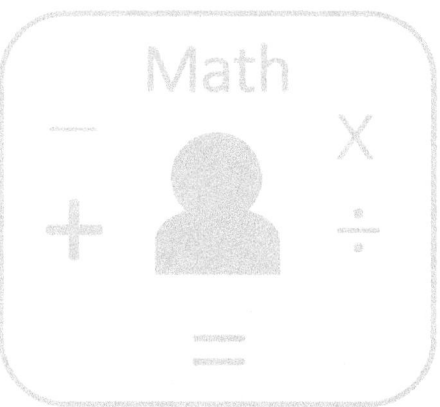

Instructions and Tips:

- You have 90 minutes to complete this worksheet
- This worksheet consists of 4 questions
- Write answers in the spaces provided
- All working must be clearly shown

Student Name: _____

Student ID: _____

Date: __ /__ /____

Total Score:

Highest Score:

Tutor's Comments:

Access more free worksheets at www.868tutors.com

Question 1

You are required to solve each triangle below. Each triangle is a non-right-angled triangle. In the space provided, sketch a triangle for reference. You need to decide whether to use the sine rule or cosine rule (or both) to solve the triangle. You can relabel the triangle where appropriate.

(a) ABC is a triangle
AB = 10 m
BC = 16 m
<ABC = 110°

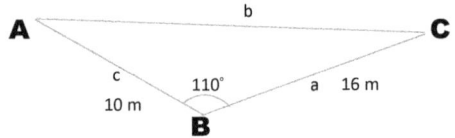

Solving for b   (Applying cosine rule)

$b^2 = a^2 + c^2 - 2ac \times \cos B$

$b^2 = (16)^2 + (10)^2 - 2(16)(10) \times \cos 110°$

$b^2 = 256 + 100 - 320 \cos 110°$

$b^2 = 256 + 100 -- 320 \cos 110°$

$b^2 = 356 + 320 \cos 110°$

$b^2 = 356 + 109.4464459$

$b^2 = 465.4464459$ **b = 21.57 m (to 2 decimal places)**

Solving for A (Applying cosine rule)

$a^2 = b^2 + c^2 - 2bc \times \cos A$

$(16)^2 = 465.4464459 + 100 - 2(21.57420789)(10) \times \cos A$

$256 - 565.4464459 = -431.4841578 \cos A$

$-309.4464459 = -431.4841578 \cos A$

$-431.4841578 \cos A = -309.4464459$

$\cos A = \dfrac{-309.4464459}{-431.4841578}$ $A = \cos^{-1}(0.717167572)$ **A = 44.18° (to 2 decimal places)**

$C = 180° - (44.18° + 110°)$ **C = 25.82° (to 2 decimal places)**

(5 marks)

(b) ABC is a triangle
AB = 17 m
BC = 9 m
< ABC = 112°

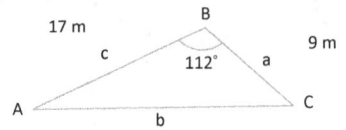

Solving for b (Applying cosine rule)

$b^2 = a^2 + c^2 - 2ac \times \cos B$

$b^2 = (9)^2 + (17)^2 - 2(9)(17) \times \cos 112°$

$b^2 = 81 + 289 - 306 \cos 112°$

$b^2 = 370 - -114.6296176$

$b^2 = 370 + 114.6296176$

$b^2 = 484.6296176$

$b = \sqrt{484.6296176}$

b = 22.01 m (to 2 decimal places)

Solving for A (applying sine rule)

$\dfrac{a}{\sin A} = \dfrac{b}{\sin B} = \dfrac{c}{\sin C}$

$\dfrac{a}{\sin A} = \dfrac{b}{\sin B}$

$\dfrac{9}{\sin A} = \dfrac{22.01}{\sin 112°}$ $22.01 \times \sin A = 9 \times \sin 112°$ $\sin A = \dfrac{9 \times \sin 112°}{22.01}$ $\sin A = \dfrac{8.344654691}{22.01}$

$\sin A = 0.379130154$
$A = \sin^{-1}(0.379130154)$

A = 22.28° (to 2 decimal places)

C = 180° - (112° + 22.28°) (Internal angles of a triangle sum to 180°)

C = 180° - 134.28°

C = 45.72° (to 2 decimal places)

(5 marks)

(c) ABC is a triangle
AB = 10 m
AC = 14 m
BC = 12.6 m

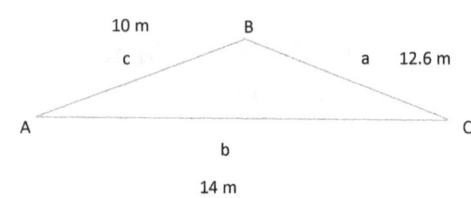

Solving for A (Applying cosine rule)

$a^2 = b^2 + c^2 - 2bc \times \cos A$

$(12.6)^2 = (14)^2 + (10)^2 - 2(14)(10) \times \cos A$

$(12.6)^2 = 196 + 100 - 280 \cos A$

$158.76 - 296 = -280 \cos A$

$-280 \cos A = -137.24$

$\cos A = \dfrac{-137.24}{-280}$

$A = \cos^{-1}\left(\dfrac{-137.24}{-280}\right)$

$A = \cos^{-1}(0.490142857)$

A = 60.65° (to 2 decimal places)

Solving for C (Applying cosine rule) $c^2 = a^2 + b^2 - 2ab \times \cos C$

$(10)^2 = (12.6)^2 + (14)^2 - 2(12.6)(14) \times \cos C =$

$100 = 158.76 + 196 - 352.8 \times \cos C$

$100 = 158.76 + 196 - 352.8 \times \cos C$

$100 = 354.76 - 352.8 \times \cos C$

$100 - 354.76 = -352.8 \times \cos C$

$-352.8 \times \cos C = -254.76$

$\cos C = \dfrac{-254.76}{-352.8}$ $\cos C = 0.722108843$

$C = \cos^{-1}(0.722108843)$

C = 43.77° (to 2 decimal places)

Solving for B (Internal angles in a triangle sum to 180°)

B = 180° - (60.65° + 43.77°)

B = 180° - (104.42°)

B = 75.58° (to 2 decimal places)

(5 marks)

(d) PQR is a triangle
PR = 6.3 cm
PQ = 6.5 cm
QR = 5.0 cm

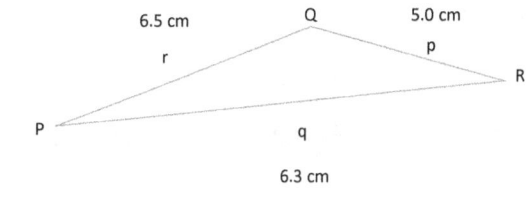

Solving for P (Applying cosine rule)

$p^2 = q^2 + r^2 - 2qr\cos P$

$(5)^2 = (6.3)^2 + (6.5)^2 - 2(6.3)(6.5)\cos P$

$25 = 39.69 + 42.25 - 81.9 \cos P$

$25 = 81.94 - 81.9 \cos P$

$25 - 81.94 = -81.9 \cos P$

$-81.9 \cos P = -56.94$

$\cos P = \dfrac{-56.94}{-81.9}$

$\cos P = 0.695238095$

$P = \cos^{-1}(0.695238095)$

P = 45.95° (to 2 decimal places)

Solving for R (Applying Cosine rule)

$r^2 = p^2 + q^2 - 2pq\cos R$

$(6.5)^2 = (5)^2 + (6.3)^2 - 2(5)(6.3)\cos R$

$42.25 = 25 + 39.69 - 63\cos R$

$42.25 = 64.69 - 63\cos R$

$-63 \cos R = 42.25 - 64.69$

$-63 \cos R = -22.44$

$\cos R = \dfrac{-22.44}{-63}$ $\cos R = 0.356190476$

$R = \cos^{-1}(0.356190476)$

R = 69.13° (to 2 decimal places)

Solving for Q (Internal angles in a triangle sum to 180°)

$Q = 180° - (69.13° + 45.95°)$

$Q = 180° - (115.08°)$

Q = 64.92° (to 2 decimal places)

(5 marks)

Question 2

Solve each triangle below. (Relabel if necessary)

(a) XY = 10 cm
<YXZ = 30°
<XYZ = 40°

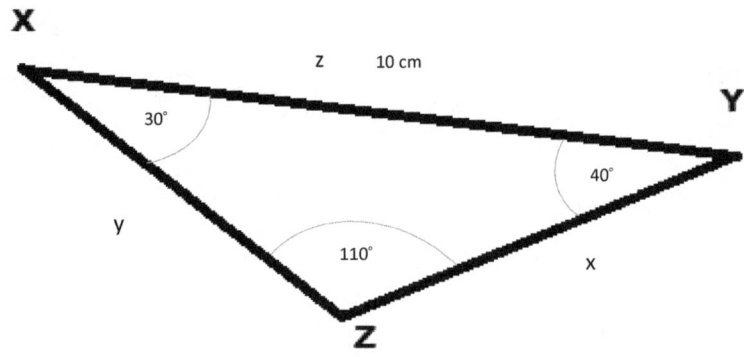

Internal angles in a triangle sum to 180°
Z = 180° - (40° + 30°)
Z = 180° - (70°)   Z = 110°

Solving for x (Applying sine rule)

$$\frac{x}{\sin x} = \frac{y}{\sin y} = \frac{z}{\sin z}$$

$\frac{x}{\sin x} = \frac{z}{\sin z}$ $\frac{x}{\sin 30°} = \frac{10}{\sin 110°}$ x sin 110° = 10 sin 30°

x sin 110° = 10 sin 30°

$x = \frac{10 \sin 30°}{\sin 110°}$ $x = \frac{5}{0.93969262}$ **x = 5.32 cm (to 2 decimal places)**

Solving for y (Applying sine rule)

$\frac{y}{\sin y} = \frac{z}{\sin z}$ $\frac{y}{\sin 40°} = \frac{10}{\sin 110°}$

y sin 110° = 10 sin 40° $y = \frac{10 \sin 40°}{\sin 110°}$ $y = \frac{6.427876097}{0.93969262}$ **y = 6.84 cm (to 2 decimal places)**

(5 marks)

(b) LM = 5 cm
LN = 7.5 cm
MN = 3 cm

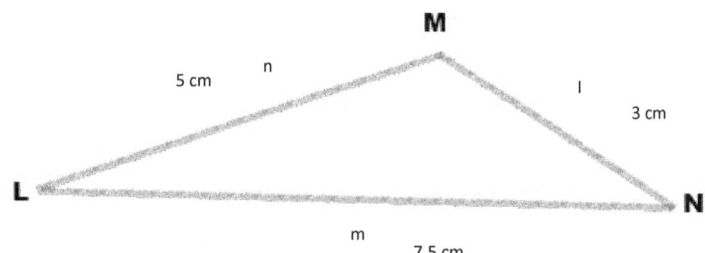

Solving for L (applying cosine rule)

$l^2 = m^2 + n^2 - 2mn\cos L$

$(3)^2 = (7.5)^2 + (5)^2 - 2(7.5)(5) \cos L$

$9 = 56.25 + 25 - 75\cos L$

$9 = 81.25 - 75\cos L$

$81.25 - 75\cos L = 9$

$-75\cos L = 9 - 81.25$

$-75\cos L = -72.25$

$\cos L = \frac{-72.25}{-75}$ $\cos L = 0.963333333$ $L = \cos^{-1}(0.963333333)$

L = 15.56° (to 2 decimal places)

Solving for N (Applying cosine rule)

$n^2 = m^2 + l^2 - 2ml\cos N$

$(5)^2 = (7.5)^2 + (3)^2 - 2(7.5)(3) \cos N$

$25 = 56.25 + 9 - 45\cos N$

$25 = 56.25 + 9 - 45\cos N$

$65.25 - 45 \cos N = 25$

$- 45 \cos N = 25 - 65.25$

$-45 \cos N = -40.25$

$\cos N = \frac{-40.25}{-45}$ $\cos N = 0.894444444$

$N = \cos^{-1}(0.894444444)$

N = 26.56° (to 2 decimal places)

Internal angles in a triangle sum to 180°

M + N + L = 180°

M = 180° - (N + L)   M = 180° - (15.56° + 26.56°) = **137.88°**

M = 137.88° (to 2 decimal places)

(5 marks)

(c) XY = 9 cm
XZ = 12 cm
YZ = 5 cm

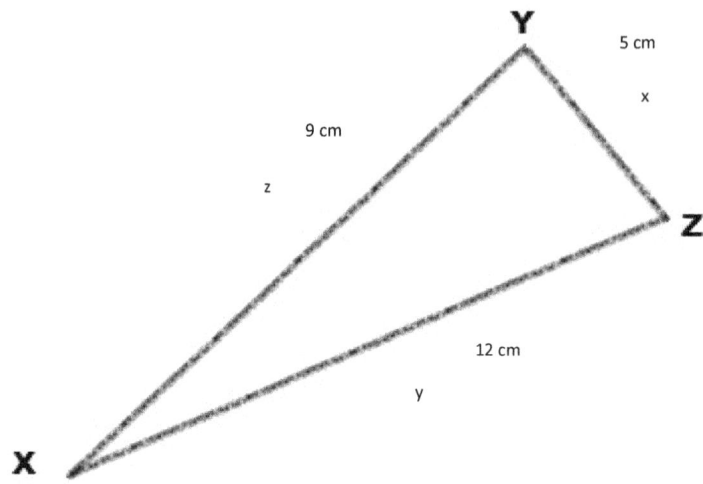

Solving for X (Applying cosine rule)

$x^2 = y^2 + z^2 - 2yz\cos X$

$(5)^2 = (12)^2 + (9)^2 - 2(12)(9)\cos X$ $(5)^2 = 144 + 81 - 2(12)(9)\cos X$

$(5)^2 = 144 + 81 - 216\cos X$ $25 = 225 - 216\cos X$

$225 - 216\cos X = 25$

$-216\cos X = 25 - 225$

$-216\cos X = -200$

$\cos X = \dfrac{-200}{-216} = 0.925925925$ $X = \cos^{-1}(0.925925925)$ **X = 22.19° (to 2 decimal places)**

Solving for Y (Applying sine rule)

$\dfrac{y}{\sin Y} = \dfrac{x}{\sin X}$

$\dfrac{12}{\sin Y} = \dfrac{5}{\sin 22.19160657}$

$5\sin Y = 12\sin 22.19160657$

$\sin Y = \dfrac{12\sin 22.19160657}{5} = 0.906492358$

$Y = \sin^{-1}(0.906492358)$ **Y = 65.03° (to 2 decimal places)**

Solving for Z (Internal angles in a triangle sum to 180°)

$Z = 180° - (65.03° + 22.19°)$

$Z = 180° - 87.22°$ **Z = 92.78° (to 2 decimal places)**

(5 marks)

Question 3

Solve each triangle. Sketches should be drawn for reference. Relabel where necessary.

(a) Triangle XYZ
    X = 75°   Y = 48°   x = 21mm

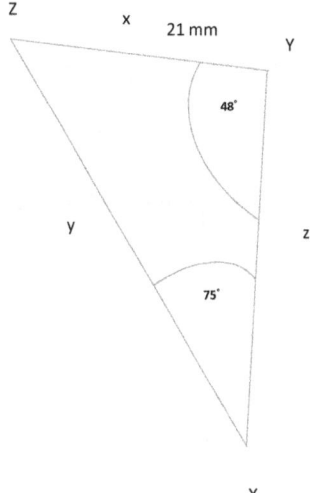

Internal angles in a triangle sum to 180°

$Z = 180° - (75° + 48°)$

$\boxed{Z = 57°}$

Solving for y (Applying sine rule)

$\frac{y}{\sin Y} = \frac{x}{\sin X}$ $\frac{y}{\sin 48°} = \frac{21}{\sin 75°}$

$y \sin 75° = 21 \sin 48°$

$y = \frac{21 \sin 48°}{\sin 75°}$ $y = \frac{15.60604134}{0.965925826}$

$\boxed{y = 16.16 \text{ mm (to 2 decimal places)}}$

Solving for z (Applying sine rule)

$\frac{z}{\sin Z} = \frac{x}{\sin X}$ $\frac{z}{\sin 57°} = \frac{21}{\sin 75°}$

$z \sin 75° = 21 \sin 57°$

$z = \frac{21 \sin 57°}{\sin 75°}$ $z = \frac{17.61208193}{\sin 75°}$

$\boxed{z = 18.23 \text{ mm (to 2 decimal places)}}$

(5 marks)

(b) **Triangle XYZ**

$X = 20°$ $Z = 102°$ $z = 11$ cm

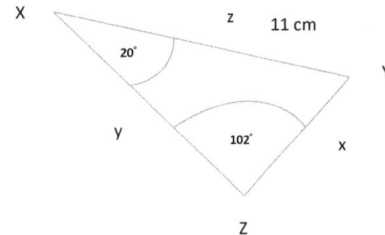

Solving for x (Applying sine rule)

$$\frac{x}{\sin X} = \frac{z}{\sin Z}$$

$$\frac{x}{\sin 20°} = \frac{11}{\sin 102°}$$

X sin 102° = 11 sin 20°

$$x = \frac{11 \sin 20°}{\sin 102°}$$

$$x = \frac{11 \sin 20°}{\sin 102°}$$

$$x = \frac{3.762221577}{0.9781476}$$ **x = 3.85 cm (to 2 decimal places)**

Solving for y (Applying sine rule)

$$\frac{y}{\sin Y} = \frac{z}{\sin Z}$$ $Y = 180° - (102° + 20°) = 58°$ (Internal angles in a triangle sum to 180°)

$$\frac{y}{\sin 58°} = \frac{11}{\sin 102°}$$

y sin 102° = 11 sin 58°

$$y = \frac{11 \sin 58°}{\sin 102°}$$

$$y = \frac{9.328529058}{0.9781476}$$

y = 9.54 cm (to 2 decimal places)

(5 marks)

(c) Triangle PQR

$p = 16$ cm $q = 14$ cm $R = 40°$

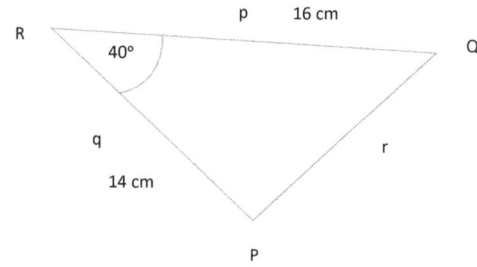

Solving for r (Applying cosine rule)

$r^2 = p^2 + q^2 - 2pq \cos R$

$r^2 = (16)^2 + (14)^2 - 2(16)(14) \cos R$

$r^2 = 256 + 196 - 2(16)(14) \cos R$

$r^2 = 452 - 448 \cos 40°$

$r^2 = 452 - 343.1879105$

$r^2 = 108.8120895$

$r = \sqrt{108.8120895}$

r = 10.43 cm (to 2 decimal places)

Solving for Q (Applying sine rule)

$\dfrac{q}{\sin Q} = \dfrac{r}{\sin R}$

$\dfrac{14}{\sin Q} = \dfrac{10.43}{\sin 40°}$

$14 \sin 40° = 10.43 \sin Q$

$10.43 \sin Q = 14 \sin 40°$

$\sin Q = \dfrac{14 \sin 40°}{10.43}$

$\sin Q = 0.86280216$

$Q = \sin^{-1}(0.86280216)$

Q = 59.63° (to 2 decimal places)

Internal angles in a triangle sum to 180°

$P = 180° - (59.63° + 40°)$

$P = 180° - 99.63°$

P = 80.37° (to 2 decimal places)

(5 marks)

Question 4

Find the area of each triangle below:

(a) Triangle ABC

$a = 105$ mm $b = 75$ mm $C = 150°$

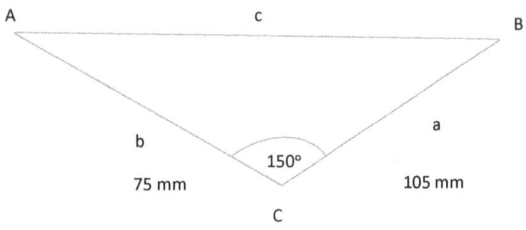

Area of triangle = $\frac{1}{2}$ ab × sinC

Area of triangle = $\frac{1}{2}(105)(75) \times \sin 150°$

Area of triangle = 3937.5×0.5

Area of triangle = 1968.75 mm² (to 2 decimal places)

(5 marks)

(b) Triangle ABC

$a = 4.5$ mm $c = 3$ mm $B = 35°$

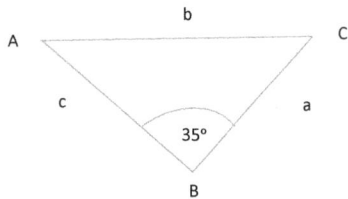

Area of triangle = $\frac{1}{2}$ ac × sinB

Area of triangle = $\frac{1}{2}(4.5)(3) \times \sin 35°$

Area of triangle = $6.75 \times \sin 35°$

Area of triangle = 3.87 mm² (to 2 decimal places)

(5 marks)

(c) Triangle PQR

p = 4 cm r = 3 cm Q = 89°

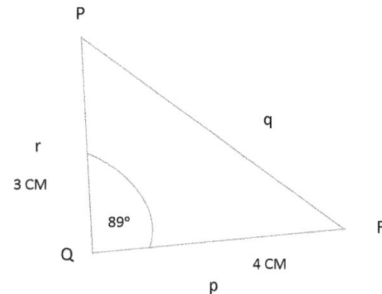

Area of triangle = $\frac{1}{2}$ pr × sinQ

Area of triangle = $\frac{1}{2}$ (4)(3) × sin 89°

Area of triangle = 6 × sin 89°

Area of triangle = 6.00 cm² (to 2 decimal places)

(5 marks)

(d) Triangle XYZ

x = 11 cm z = 7 cm Y = 87°

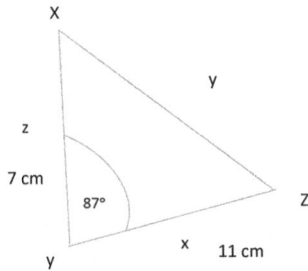

Area of triangle = $\frac{1}{2}$ xz × sinY

Area of triangle = $\frac{1}{2}$ (11)(7) × sin 87°

Area of triangle = 38.5 × sin 87°

Area of triangle = 38.45 cm² (to 2 decimal places)

(5 marks)

END OF WORKSHEET

868

868TUTORS

Preparation for

High School Mathematics

Trigonometry

(Combined)

Solutions

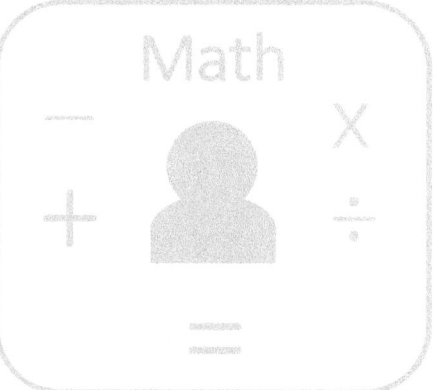

Instructions and Tips:

- ✓ You have 75 minutes to complete this worksheet
- ✓ This worksheet consists of 3 questions
- ✓ Write answers in the spaces provided
- ✓ All working must be clearly shown

Student Name: _____

Student ID: _____

Date: __/__/____

Total Score:

Highest Score:

Tutor's Comments:

Access more free worksheets at www.868tutors.com

Question 1

The illustration below (not drawn to scale) shows the position of three Form 5 Mathematics students and a flagpole, F. The students and flagpole are on a horizontal surface. April, A is standing at the base of a vertical flagpole. Samantha, S is located due south of April. Rajiv, R is located east of April. The flagpole has a height of 30 m. The angle of elevation of the top of the flagpole from Rajiv's location is 30°. The distance between Samantha and April is 42 m.

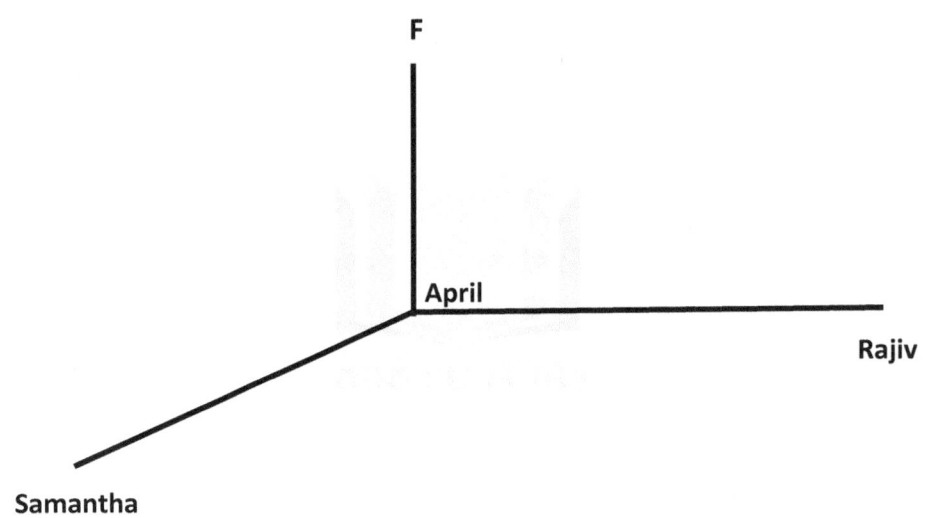

(a) Sketch separate diagrams of the triangles *RAF*, *FAS* and *SAR*. Clearly indicate on each diagram, the lengths of the given sides and angles.

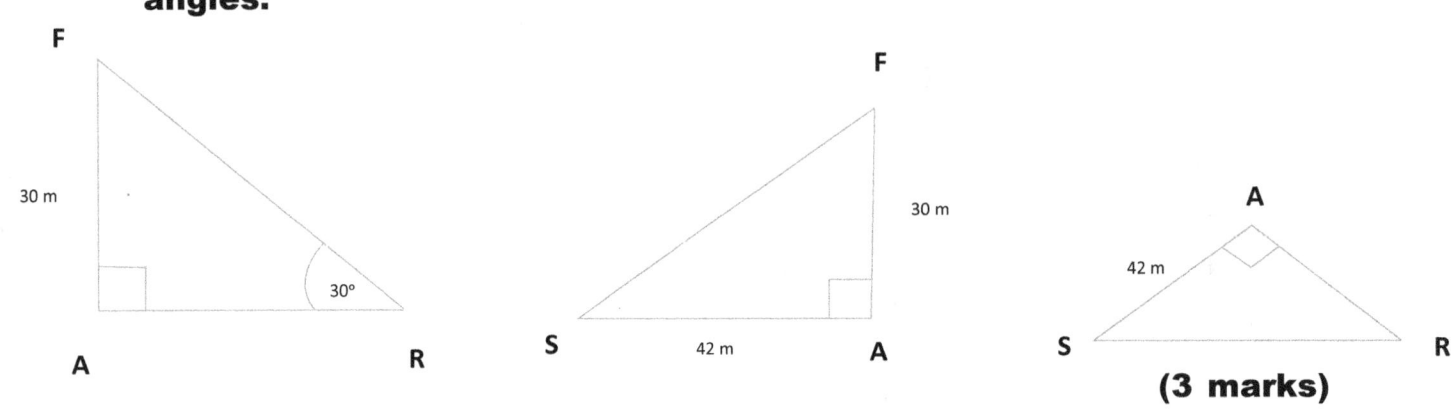

(3 marks)

(b) Calculate the straight line distance between Rajiv and April (RA) to 2 decimal places.

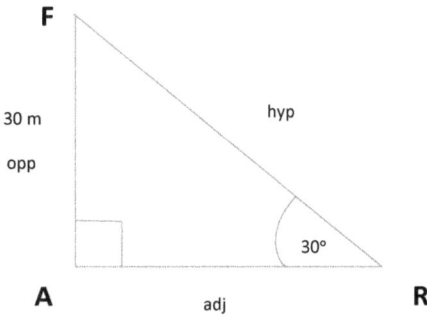

Calculating AR

Let angle FRA = y

$\tan y = \frac{\text{opposite}}{\text{adjacent}}$

$\tan 30° = \frac{30}{\text{adj}}$

adj × tan 30° = 30

adj = $\frac{30}{\tan 30°}$ adj = 51.96 m

The distance between Rajiv and April is = 51.96 m (to 2 decimal places)

(2 marks)

(c) Calculate the straight line distance between Samantha and Rajiv (SR) to 2 decimal places.

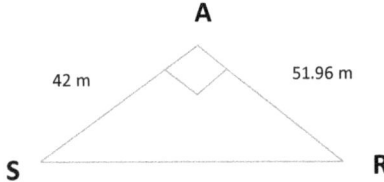

Recall that AR = 51.96 m

Apply Pythagoras' theorem

$(AS)^2 + (AR)^2 = (SR)^2$

$(SR)^2 = (AS)^2 + (AR)^2$

$(SR)^2 = (42)^2 + (51.96)^2$

$(SR)^2 = 1764 + 2699.8416$

$(SR)^2 = 4463.8416$

$(SR) = 66.81\ m$

The distance between Samantha and Rajiv = 66.81 m (to 2 decimal places)

(1 mark)

(d) Calculate the angle of elevation from Samantha's position (from ground) to the top of the flagpole to 2 decimal places. (Neglect Samantha's height)

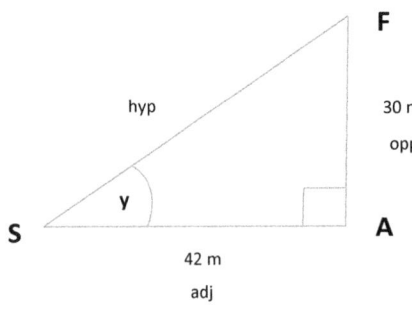

Let angle FSA = y   Solving for y

$\tan y = \frac{opp}{adj}$ $\tan y = \frac{30\ m}{42\ m}$ $\tan y = 0.714285714$ $y = \tan^{-1}(0.714285714)$

$y = 35.54°$ **angle of elevation = 35.54° (to 2 decimal places)**

(2 marks)

Question 2

In the illustration below, not drawn to scale, AB = 9.2 m, AC = 12.1 m, CD = 13.4 m, <BAC = 60° and <CBD = 40°

(a) Label the given sides and angles

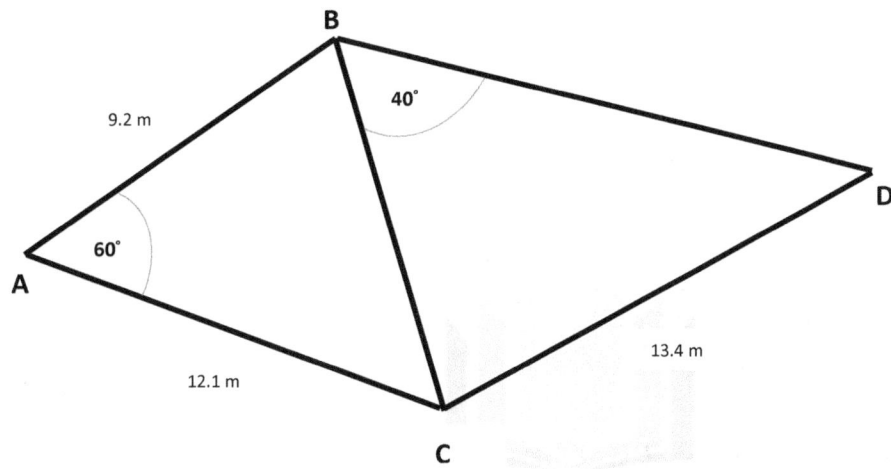

(2 marks)

Determine the following:

(b) the length of BC to 2 decimal places

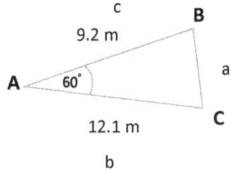

Using cosine rule

$a^2 = b^2 + c^2 - 2bc \cdot \cos A$

$a^2 = (12.1)^2 + (9.2)^2 - 2(12.1)(9.2) \cos 60°$

$a^2 = (12.1)^2 + (9.2)^2 - 2(12.1)(9.2) \cos 60°$

$a^2 = 146.41 + 84.64 - 222.64 \cos 60°$

$a^2 = 231.05 - 111.32$

$a^2 = 119.73 \quad a = 10.94 \text{ m}$ | BC = 10.94 m (to 2 decimal places) |

(2 marks)

(c) the size of <BDC to 2 decimal places

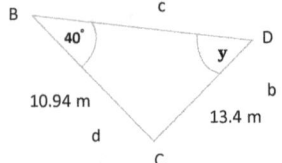

Recall BC = 10.94 m

Applying sine rule

$\dfrac{c}{\sin C} = \dfrac{d}{\sin D} = \dfrac{b}{\sin B}$

$\dfrac{10.94}{\sin D} = \dfrac{13.4}{\sin 40°}$

$10.94 \times \sin 40° = 13.4 \times \sin D$

$13.4 \times \sin D = 10.94 \times \sin 40°$

$\sin D = \dfrac{10.94 \times \sin 40°}{13.4}$

$\sin D = 0.524783317$

$D = \sin^{-1}(0.524783317)$

$D = \sin^{-1}(0.524783317)$

<BDC = 31.65° (to 2 decimal places)

(2 marks)

(d) the area of triangle ABC to 2 decimal places

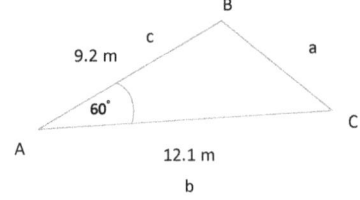

Area of triangle $= \tfrac{1}{2} bc \times \sin A$

Area of triangle $= \tfrac{1}{2}(12.1) \times (9.2) \times \sin 60°$

Area of triangle $= 55.66 \times \sin 60°$

Area of triangle ABC = 48.20 m² (to 2 decimal places)

(2 marks)

(e) the perpendicular distance from B to AC to 2 decimal places

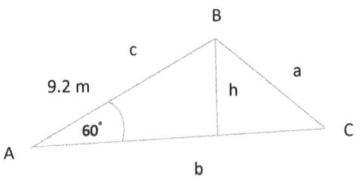

Area of triangle $= \dfrac{b \times h}{2}$ *recall Area = 48.20 m²* $\dfrac{48.20}{1} = \dfrac{12.1 \times h}{2}$

$12.1 \times h = 96.40$

$h = \dfrac{96.40}{12.1}$ $h = 7.97\ m$

Perpendicular distance from B to AC = 7.97 m (to 2 decimal places)

(1 mark)

Question 3

Off the coast of Icacos Village, Cedros in Trinidad lies a submerged lighthouse. A man swims with snorkeling equipment. He is located 10 metres away in a horizontal distance from the light house. The angle of depression from the swimmer to the top of the light house is 30°. The illustration below represents the situation. We are assuming that the seabed is a horizontal surface and that the water is level.

In the illustration below, draw a line and label the angle of depression.

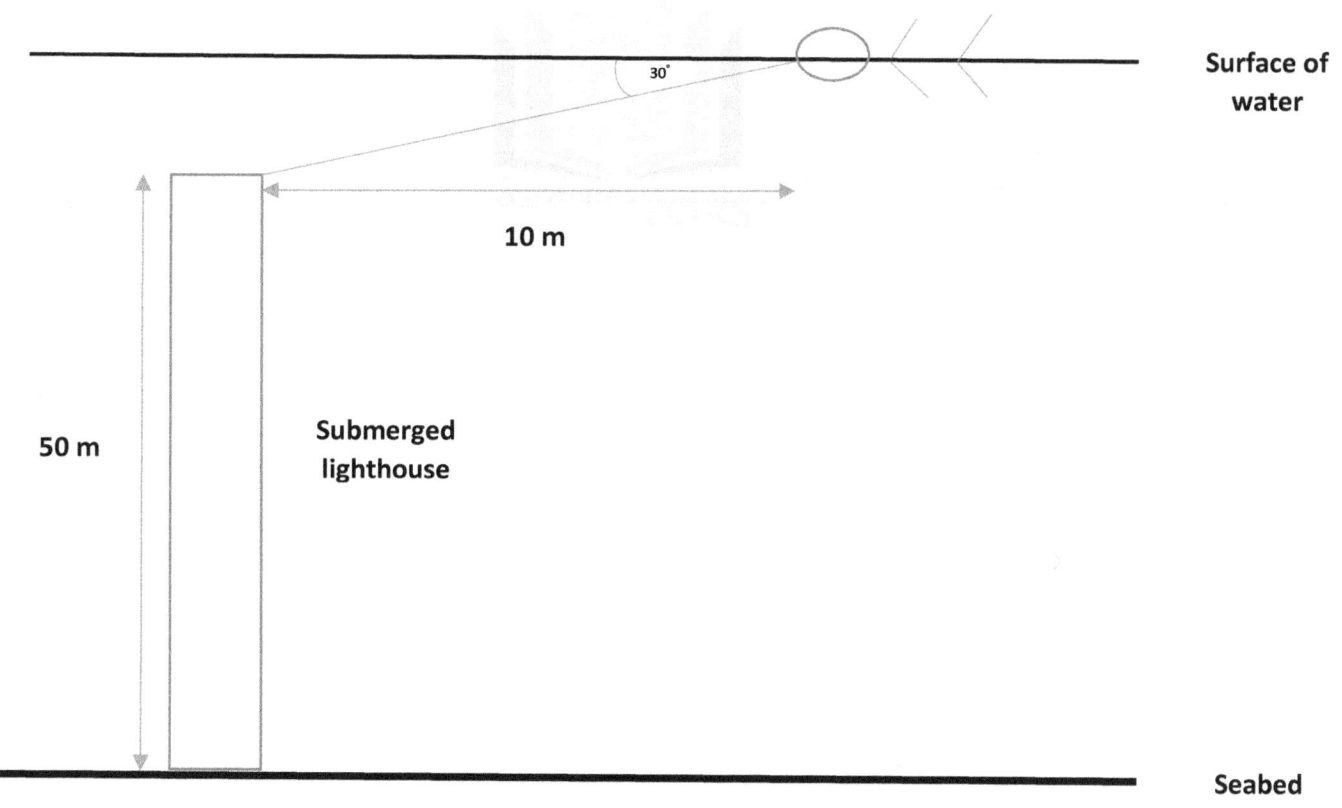

(1 mark)

The height of the lighthouse is 50m. Calculate the depth of the water in which the man swims.

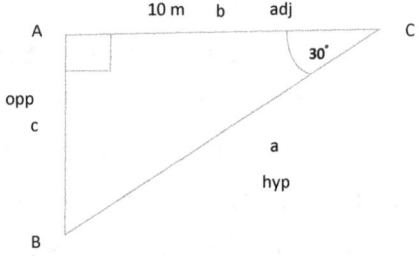

Utilizing triangle from diagram

Solving for AB

$\tan 30° = \frac{opp}{adj}$

$\tan 30° = \frac{opp}{10}$

10 × tan 30° = opp

opp = 5.77 m

Depth of water = 5.77 m + 50 m

Depth of water = 55.77 m (to 2 decimal places)

(4 marks)

END OF WORKSHEET

868

TUTORS

Preparation for

High School Mathematics

Vectors

Solutions

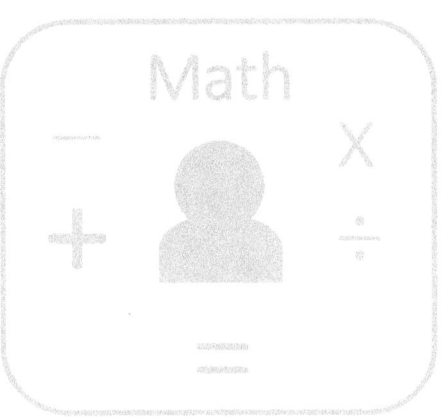

Instructions and Tips:

- You have 90 minutes to complete this worksheet
- This worksheet consists of 12 questions
- Write answers in the spaces provided
- Show all Vector Algebra
- Give your answers in the simplest form

Student Name: _____

Student ID: _____

Date: __/__/____

Total Score:

Highest Score:

Tutor's Comments:

Access more free worksheets at www.868tutors.com

Question 1

Consider the triangle XYZ.

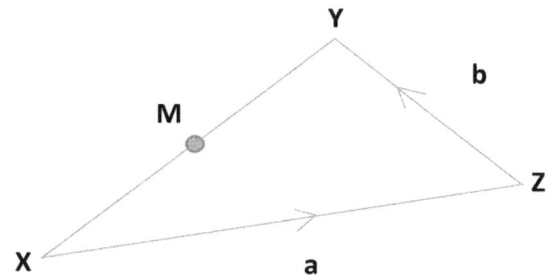

(Diagram not drawn to scale)

M is the midpoint of XY.

$\vec{XZ} = \mathbf{a}$

$\vec{ZY} = \mathbf{b}$

(a) Express $\vec{YX}$ in terms of a and b.

$\vec{YX} = \vec{YZ} + \vec{ZX}$ *(triangle law)*

$\vec{YX} = -b + -a$

$\boxed{\vec{YX} = -\mathbf{b} - \mathbf{a}}$

(2 marks)

(b) Express $\vec{YM}$ in terms of a and b.

$\vec{YM} = \frac{1}{2}\vec{YX}$

$\boxed{\vec{YM} = \frac{1}{2}(-\mathbf{b} - \mathbf{a})}$

(1 mark)

(c) Express $\vec{ZM}$ in terms of a and b.

$\vec{ZM} = \vec{ZY} + \vec{YM}$ *(triangle law)* $\vec{ZM} = b + \frac{1}{2}(-b - a)$

$\vec{ZM} = b - \frac{1}{2}b - \frac{1}{2}a$ $\vec{ZM} = \frac{1}{2}b - \frac{1}{2}a$ $\boxed{\vec{ZM} = \frac{1}{2}(\mathbf{b} - \mathbf{a})}$

(2 marks)

Question 2

Consider the triangle ABC.

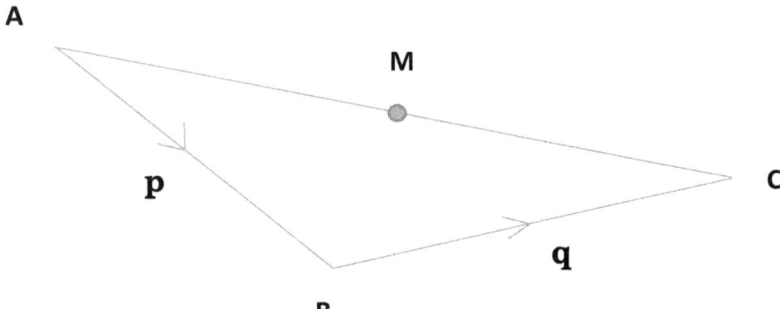

(Diagram not drawn to scale)

M is the midpoint of AC.

$\overrightarrow{AB} = \mathbf{p}$

$\overrightarrow{BC} = \mathbf{q}$

(a) Express $\overrightarrow{AC}$ in terms of p and q.

$\overrightarrow{AC} = \overrightarrow{AB} + \overrightarrow{BC}$ *(triangle law)*

$\boxed{\overrightarrow{AC} = \mathbf{p} + \mathbf{q}}$

(2 marks)

(b) Express $\overrightarrow{AM}$ in terms of p and q.

$\overrightarrow{AM} = \frac{1}{2}\overrightarrow{AC}$

$\boxed{\overrightarrow{AM} = \frac{1}{2}(\mathbf{p} + \mathbf{q})}$

(1 mark)

(c) Express $\overrightarrow{BM}$ in terms of p and q.

$\overrightarrow{BM} = \overrightarrow{BA} + \overrightarrow{AM}$ *(triangle law)* $\overrightarrow{BM} = -\mathbf{p} + \frac{1}{2}(\mathbf{p}+\mathbf{q})$ $\overrightarrow{BM} = -\mathbf{p} + \frac{1}{2}\mathbf{p} + \frac{1}{2}\mathbf{q}$

$\overrightarrow{BM} = -\frac{1}{2}\mathbf{p} + \frac{1}{2}\mathbf{q}$ $\boxed{\overrightarrow{BM} = \frac{1}{2}(\mathbf{q} - \mathbf{p})}$

(2 marks)

Question 3

Consider the triangle PQR.

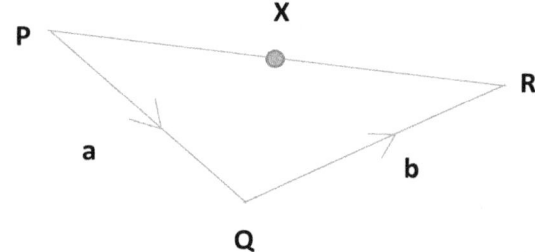

(Diagram not drawn to scale)

X is the midpoint of PR.

$\overrightarrow{PQ}$ = **a**

$\overrightarrow{QR}$ = **b**

(a) Express $\overrightarrow{PR}$ in terms of a and b.

$\overrightarrow{PR} = \overrightarrow{PQ} + \overrightarrow{QR}$ *(triangle law)*

$\boxed{\overrightarrow{PR} = \mathbf{a} + \mathbf{b}}$

(2 marks)

(b) Express $\overrightarrow{PX}$ in terms of a and b.

$\overrightarrow{PX} = \frac{1}{2}\overrightarrow{PR}$

$\boxed{\overrightarrow{PX} = \frac{1}{2}(\mathbf{a} + \mathbf{b})}$

(1 mark)

(c) Express $\overrightarrow{QX}$ in terms of a and b.

$\overrightarrow{QX} = \overrightarrow{QP} + \overrightarrow{PX}$ *(triangle law)*

$\overrightarrow{QX} = -\mathbf{a} + \frac{1}{2}(\mathbf{a}+\mathbf{b})$ $\overrightarrow{QX} = -\mathbf{a} + \frac{1}{2}\mathbf{a} + \frac{1}{2}\mathbf{b}$

$\overrightarrow{QX} = -\frac{1}{2}\mathbf{a} + \frac{1}{2}\mathbf{b}$ $\boxed{\overrightarrow{QX} = \frac{1}{2}(\mathbf{b} - \mathbf{a})}$

(2 marks)

Question 4

Consider the triangle ABC.

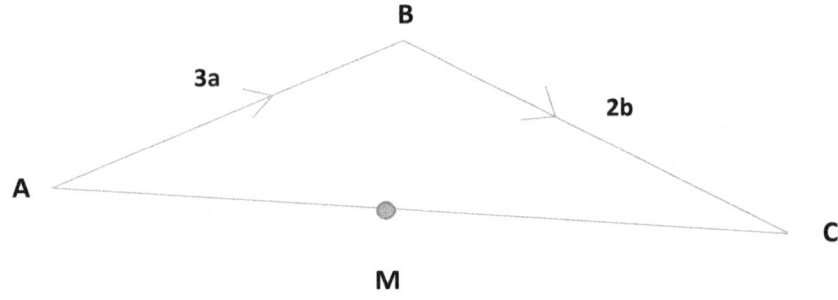

(Diagram not drawn to scale)

M is the midpoint of AC.

$\overrightarrow{AB} = 3a$

$\overrightarrow{BC} = 2b$

(a) Express $\overrightarrow{AC}$ in terms of a and b.

$\overrightarrow{AC} = \overrightarrow{AB} + \overrightarrow{BC}$ *(triangle law)*

$\boxed{\overrightarrow{AC} = 3a + 2b}$

(2 marks)

(b) Express $\overrightarrow{AM}$ in terms of a and b.

$\overrightarrow{AM} = \frac{1}{2}\overrightarrow{AC}$

$\boxed{\overrightarrow{AM} = \frac{1}{2}(3a + 2b)}$

(1 mark)

(c) Express $\overrightarrow{BM}$ in terms of a and b.

$\overrightarrow{BM} = \overrightarrow{BA} + \overrightarrow{AM}$ *(triangle law)*

$\overrightarrow{BM} = -3a + \frac{1}{2}(3a + 2b)$ $\overrightarrow{BM} = -3a + \frac{3}{2}a + 1b$

$\overrightarrow{BM} = -3a + \frac{3}{2}a + 1b$ $\boxed{\overrightarrow{BM} = b - \frac{3}{2}a}$

(2 marks)

Question 5

Consider the triangle ABC.

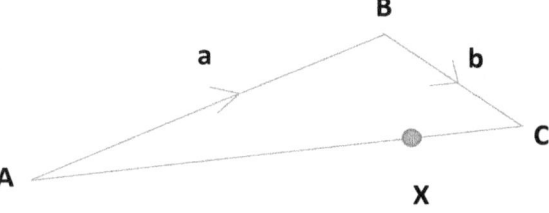

(Diagram not drawn to scale)

$\vec{AB} = \mathbf{a}$

$\vec{BC} = \mathbf{b}$

X is a point on the line AC such that AX:XC = 2:1

(a) Express $\vec{AC}$ in terms of a and b.

$\vec{AC} = \vec{AB} + \vec{BC}$ *(triangle law)*

$\boxed{\vec{AC} = \mathbf{a} + \mathbf{b}}$

(2 marks)

(b) Express $\vec{AX}$ in terms of a and b.

$\vec{AX} = \frac{2}{3}\vec{AC}$ $\boxed{\vec{AX} = \frac{2}{3}(\mathbf{a}+\mathbf{b})}$

(1 mark)

(c) Express $\vec{BX}$ in terms of a and b.

$\vec{BX} = \vec{BA} + \vec{AX}$ $\vec{BX} = -\mathbf{a} + \frac{2}{3}\mathbf{a} + \frac{2}{3}\mathbf{b}$

$\vec{BX} = -\frac{1}{3}\mathbf{a} + \frac{2}{3}\mathbf{b}$ $\boxed{\vec{BX} = \frac{1}{3}(2\mathbf{b} - \mathbf{a})}$

(2 marks)

Question 6

Consider the triangle XYZ.

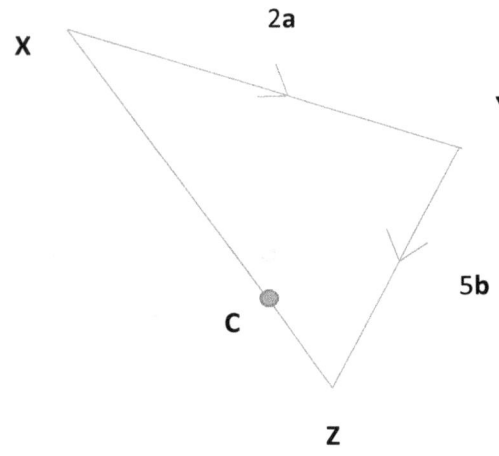

(Diagram not drawn to scale)

$\overrightarrow{XY} = 2\mathbf{a} \quad \overrightarrow{YZ} = 5\mathbf{b}$

C is a point on the line XZ such that XC: CZ = 3:1.

(a) Express $\overrightarrow{XZ}$ in terms of a and b.

$\overrightarrow{XZ} = \overrightarrow{XY} + \overrightarrow{YZ}$ *(triangle law)*

$\boxed{\overrightarrow{XZ} = 2\mathbf{a} + 5\mathbf{b}}$

(2 marks)

(b) Express $\overrightarrow{XC}$ in terms of a and b.

$XC: CZ = 3:1 \quad \overrightarrow{XC} = \frac{3}{4}\overrightarrow{XZ}$

$\boxed{\overrightarrow{XC} = \frac{3}{4}(2\mathbf{a} + 5\mathbf{b})}$

(2 marks)

(c) Show that $\overrightarrow{XC}$ is parallel to the vector 2a + 5b.

$\overrightarrow{XC} = \frac{3}{4}(2\mathbf{a} + 5\mathbf{b}) \quad \overrightarrow{XC}$ *is parallel to the vector 2a + 5b because $\overrightarrow{XC}$ contains the vector 2a + 5b*

(2 marks)

Question 7

Consider the triangle RST

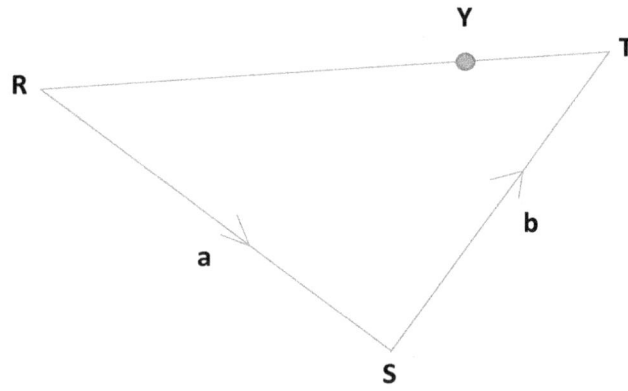

(Diagram not drawn to scale)

$\vec{RS} = \mathbf{a}$

$\vec{ST} = \mathbf{b}$

Y is a point on the line RT such that RY: YT = 5:1.

(a) Express $\vec{RT}$ in terms of a and b.

$\vec{RT} = \vec{RS} + \vec{ST}$ *(triangle law)*

$\boxed{\vec{RT} = \mathbf{a} + \mathbf{b}}$

(2 marks)

(b) Express $\vec{RY}$ in terms of a and b.

RY: YT = 5:1 $\vec{RY} = \frac{5}{6}\vec{RT}$

$\boxed{\vec{RY} = \frac{5}{6}(\mathbf{a} + \mathbf{b})}$

(2 marks)

(c) Express $\vec{SY}$ in terms of a and b.

$\vec{SY} = \vec{SR} + \vec{RY}$ *(triangle law)* $\vec{SY} = -\mathbf{a} + \frac{5}{6}(\mathbf{a}+\mathbf{b})$ $\vec{SY} = -\mathbf{a} + \frac{5}{6}\mathbf{a} + \frac{5}{6}\mathbf{b}$

$\vec{SY} = -\frac{1}{6}\mathbf{a} + \frac{5}{6}\mathbf{b}$ $\boxed{\vec{SY} = \frac{1}{6}(5\mathbf{b} - \mathbf{a})}$

(2 marks)

Question 8

Consider the parallelogram PQRS.

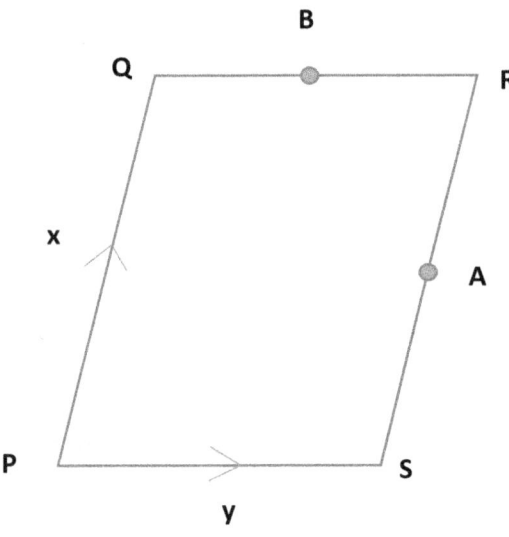

(Diagram not drawn to scale)

$\overrightarrow{PQ} = x$ $\overrightarrow{PS} = y$

A is the midpoint of the line SR.   B is the midpoint of the line QR.

(a) Express $\overrightarrow{SA}$ in terms of x.

$\overrightarrow{SA} = \frac{1}{2} \overrightarrow{SR}$ $\overrightarrow{SR} = \overrightarrow{PQ}$ *(equal vectors)* $\overrightarrow{SR} = x$

$\overrightarrow{SA} = \frac{1}{2} \overrightarrow{SR}$ $\boxed{\overrightarrow{SA} = \frac{1}{2} x}$

(b) Express $\overrightarrow{QB}$ in terms of y.

$\overrightarrow{QB} = \frac{1}{2} \overrightarrow{QR}$ $\overrightarrow{QR} = \overrightarrow{PS}$ *(equal vectors)* $\overrightarrow{QR} = y$

$\boxed{\overrightarrow{QB} = \frac{1}{2} y}$

(c) Express $\overrightarrow{SQ}$ in terms of x and y.

$\overrightarrow{SQ} = \overrightarrow{SP} + \overrightarrow{PQ}$ *(triangle law)*

$\overrightarrow{SQ} = -y + x$ $\boxed{\overrightarrow{SQ} = x - y}$

(4 marks)

(d) Express $\vec{AB}$ in terms of x and y.

$\vec{AB} = \vec{AR} + \vec{RB}$

$\vec{AR} = \frac{1}{2}x$ $\qquad\qquad$ $\vec{AR} = \vec{SA}$

$\vec{RB} = -\frac{1}{2}y$ $\qquad\qquad$ $\vec{RB} = \frac{1}{2}\vec{RQ}$

$\vec{AB} = \frac{1}{2}x + -\frac{1}{2}y$ $\qquad$ $\vec{RQ} = -\vec{PS}$ $\quad$ (inverse vectors)

$\vec{AB} = \frac{1}{2}x - \frac{1}{2}y$ $\qquad\qquad$ $\vec{PS} = y$

$\boxed{\vec{AB} = \frac{1}{2}(x - y)}$

(e) Show that $\vec{SQ}$ is parallel to $\vec{AB}$.

Recall $\vec{SQ} = x - y$

Recall $\vec{AB} = \frac{1}{2}(x - y)$

$\vec{SQ}$ is parallel to $\vec{AB}$ because they both contain the same vector $x - y$.

(4 marks)

Question 9

Consider the parallelogram ABCD.

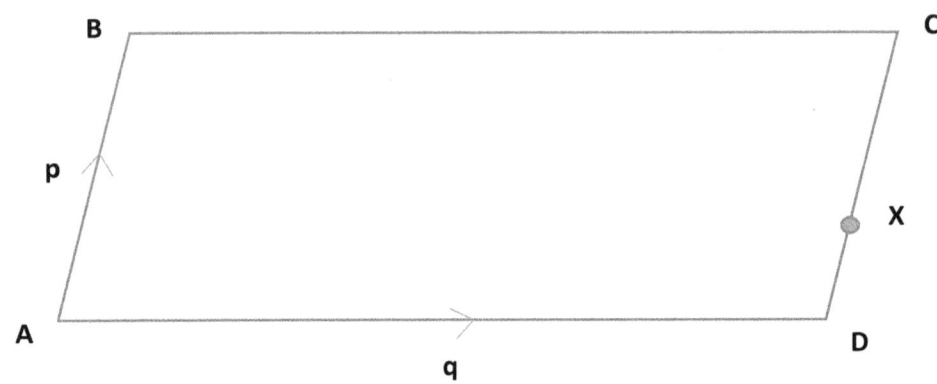

(Diagram not drawn to scale)

$\overrightarrow{AB} = \mathbf{p}$

$\overrightarrow{AD} = \mathbf{q}$

X is a point on DC such that DX:XC = 1:5.

(a) Express $\overrightarrow{DX}$ in terms of p.

DX:XC = 1:5

$\overrightarrow{DX} = \frac{1}{6}\overrightarrow{DC}$ $\quad \overrightarrow{DC} = \overrightarrow{AB} = p$ *(equal vectors)*

$\boxed{\overrightarrow{DX} = \frac{1}{6}\mathbf{p}}$

(1 mark)

(b) Express $\overrightarrow{XC}$ in terms of p.

$\overrightarrow{XC} = \frac{5}{6}\overrightarrow{DC}$ $\quad \overrightarrow{DC} = \overrightarrow{AB} = p$ *(equal vectors)*

$\boxed{\overrightarrow{XC} = \frac{5}{6}\mathbf{p}}$

(2 marks)

(c) Express $\overrightarrow{CB}$ in terms of q.

$\overrightarrow{CB} = -\overrightarrow{AD}$ $\overrightarrow{CB} = -\overrightarrow{BC}$

$\boxed{\overrightarrow{CB} = -q}$ $\overrightarrow{BC} = \overrightarrow{AD}$

(1 mark)

(d) Express $\overrightarrow{XB}$ in terms of p and q.

$\overrightarrow{XB} = \overrightarrow{XC} + \overrightarrow{CB}$

Recall $\overrightarrow{XC} = \frac{5}{6}p$

Recall $\overrightarrow{CB} = -q$

$\boxed{\overrightarrow{XB} = \frac{5}{6}p - q}$

(2 marks)

Question 10

The points X, Y and Z have coordinates as follows: X (1, 4) Y (-6,-5) and Z (1,-7).

Express each of the following in the form: $\begin{pmatrix} x \\ y \end{pmatrix}$

(a) $\overrightarrow{OX} = \begin{pmatrix} 1 \\ 4 \end{pmatrix}$

(b) $\overrightarrow{OY} = \begin{pmatrix} -6 \\ -5 \end{pmatrix}$

(c) $\overrightarrow{OZ} = \begin{pmatrix} 1 \\ -7 \end{pmatrix}$

(3 marks)

Determine the value of the following:

(d) $|\overrightarrow{OX}| = \sqrt{x^2 + y^2} = \sqrt{(1)^2 + (4)^2} = \boxed{\sqrt{17}}$ units

(e) $|\overrightarrow{OY}| = \sqrt{x^2 + y^2} = \sqrt{(-6)^2 + (-5)^2} = \boxed{\sqrt{61}}$ units

(f) $|\overrightarrow{OZ}| = \sqrt{x^2 + y^2} = \sqrt{(1)^2 + (-7)^2} = \boxed{\sqrt{50}}$ units

(3 marks)

(g) Draw a diagram to illustrate the vectors $\overrightarrow{OX}$ $\overrightarrow{OY}$ and $\overrightarrow{OZ}$.

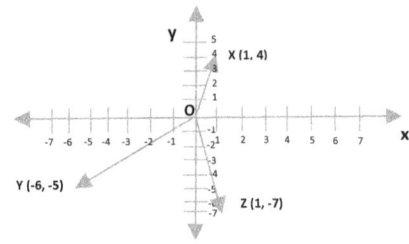

(4 marks)

Question 11

Consider the diagram below:

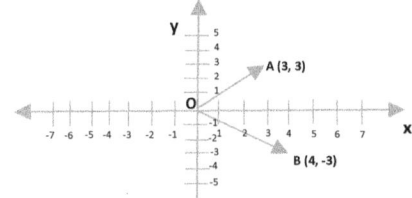

The coordinates of A and B are given as A (3, 3) and B (4,-3).

Express each of the following in the form $\begin{pmatrix} x \\ y \end{pmatrix}$

(a) $\overrightarrow{OA} = \begin{pmatrix} 3 \\ 3 \end{pmatrix}$

(b) $\overrightarrow{OB} = \begin{pmatrix} 4 \\ -3 \end{pmatrix}$

(c) $\overrightarrow{AB} = \overrightarrow{AO} + \overrightarrow{OB}$

$\overrightarrow{AB} = \begin{pmatrix} -3 \\ -3 \end{pmatrix} + \begin{pmatrix} 4 \\ -3 \end{pmatrix} = \begin{pmatrix} 1 \\ -6 \end{pmatrix}$ $\boxed{\overrightarrow{AB} = \begin{pmatrix} 1 \\ -6 \end{pmatrix}}$ (3 marks)

(d) Given that $\overrightarrow{OA} = \overrightarrow{CB}$, determine the coordinates of the point C.

To determine the coordinates of the point C, we need to first determine $\overrightarrow{OC}$

$\overrightarrow{OA} = \overrightarrow{CB}$

Therefore, $\overrightarrow{CB} = \begin{pmatrix} 3 \\ 3 \end{pmatrix}$

$\overrightarrow{CO} + \overrightarrow{OB} = \overrightarrow{CB}$

$\begin{pmatrix} x \\ y \end{pmatrix} + \begin{pmatrix} 4 \\ -3 \end{pmatrix} = \begin{pmatrix} 3 \\ 3 \end{pmatrix}$ $x + 4 = 3$ $x = 3 - 4 = -1$

$y + -3 = 3$ $y = 3 + 3 = 6$

$\overrightarrow{CO} = \begin{pmatrix} -1 \\ 6 \end{pmatrix}$ $\overrightarrow{OC} = \begin{pmatrix} 1 \\ -6 \end{pmatrix}$ $\boxed{C = (1, -6)}$ (4 marks)

Question 12

The points A, B and C have coordinates as follows: A (5, 2) B (-1, 5) and C (-4,-3).

Express each of the following in the form : $\begin{pmatrix} x \\ y \end{pmatrix}$

(a) $\vec{OA} = \begin{pmatrix} 5 \\ 2 \end{pmatrix}$

(b) $\vec{OB} = \begin{pmatrix} -1 \\ 5 \end{pmatrix}$

(c) $\vec{OC} = \begin{pmatrix} -4 \\ -3 \end{pmatrix}$

(d) $\vec{AB} = \vec{AO} + \vec{OB}$

$\vec{AB} = \begin{pmatrix} -5 \\ -2 \end{pmatrix} + \begin{pmatrix} -1 \\ 5 \end{pmatrix} = \begin{pmatrix} -6 \\ 3 \end{pmatrix}$ $\boxed{\vec{AB} = \begin{pmatrix} -6 \\ 3 \end{pmatrix}}$

(e) $\vec{AC} = \vec{AO} + \vec{OC}$

$\vec{AC} = \begin{pmatrix} -5 \\ -2 \end{pmatrix} + \begin{pmatrix} -4 \\ -3 \end{pmatrix} = \begin{pmatrix} -9 \\ -5 \end{pmatrix}$ $\boxed{\vec{AC} = \begin{pmatrix} -9 \\ -5 \end{pmatrix}}$

(f) $\vec{BC} = \vec{BO} + \vec{OC}$

$\vec{BC} = \begin{pmatrix} 1 \\ -5 \end{pmatrix} + \begin{pmatrix} -4 \\ -3 \end{pmatrix} = \begin{pmatrix} -3 \\ -8 \end{pmatrix}$ $\boxed{\vec{BC} = \begin{pmatrix} -3 \\ -8 \end{pmatrix}}$

(5 marks)

(g) **Draw a diagram to illustrate the vectors $\vec{OA}$, $\vec{OB}$ and $\vec{OC}$.**

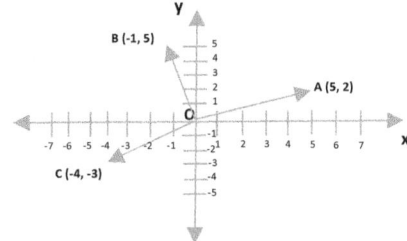

(4 marks)

(h) **Determine the values of the following:**

$|\vec{OA}| = \sqrt{x^2 + y^2} = \sqrt{(5)^2 + (2)^2} = \boxed{\sqrt{29} \text{ units}}$

$|\vec{OB}| = \sqrt{x^2 + y^2} = \sqrt{(-1)^2 + (5)^2} = \boxed{\sqrt{26} \text{ units}}$

$|\vec{OC}| = \sqrt{x^2 + y^2} = \sqrt{(-4)^2 + (-3)^2} = \sqrt{25} = \boxed{5 \text{ units}}$

(3 marks)

END OF WORKSHEET

About the Author

Anim Adrian Amarsingh was born on November, 20th 1989 in St Clair, Port of Spain, Trinidad and Tobago. In 2011, he graduated with a Bachelor of Science (Summa Cum Laude) in Electrical Engineering from Morgan State University. In 2013, he graduated with a Master of Science in Electrical and Computer Engineering from The University of Florida. He is the creator of 868 Tutors, an online educational services provider.

www.ingramcontent.com/pod-product-compliance
Lightning Source LLC
Chambersburg PA
CBHW080650190526
45169CB00006B/2051